Abdallah Ouerdane

Optics II

Abdallah Ouerdane

Optics II

Wave optics course and corrected exercises

ScienciaScripts

Imprint

Any brand names and product names mentioned in this book are subject to trademark, brand or patent protection and are trademarks or registered trademarks of their respective holders. The use of brand names, product names, common names, trade names, product descriptions etc. even without a particular marking in this work is in no way to be construed to mean that such names may be regarded as unrestricted in respect of trademark and brand protection legislation and could thus be used by anyone.

Cover image: www.ingimage.com

This book is a translation from the original published under ISBN 978-620-3-44958-7.

Publisher:
Sciencia Scripts
is a trademark of
Dodo Books Indian Ocean Ltd. and OmniScriptum S.R.L publishing group

120 High Road, East Finchley, London, N2 9ED, United Kingdom
Str. Armeneasca 28/1, office 1, Chisinau MD-2012, Republic of Moldova, Europe
Printed at: see last page
ISBN: 978-620-5-76225-7

I. Preamble

Light is the origin of several mathematical subjects such as analytic functions functions with complex variables surface and volume integrals. Light is also present in physics such as geometric optics, wave optics ͻelectromagnetismrelativistic and quantum electromagnetic waves. The photon elementary quantity of light has taken the place of the electron by multiplying telephone communications in number of subscribers and quality of services for information security and fast management by ensuring a high speed in internet.

Light has always been a subject of current interest during the e cennies and even centuries going back to the Greek scholars like Pythagoras, Aristotle and Democritus quoting only these three and there are others. The scholar who translated the physical sciences from the Greek cavillation and brought his knowledge and know-how would be Ibn El Heithem in the seventh century of our era who gave optics itself a vertiginous advance. It was not until the sixteenth century that light was reviewed and studied in the manner known today by introducing laws. Great progress was made by scientists like Newton who was at the origin of classical mechanics and who also gave his interpretation of optics as a corpuscular phenomenon. Grimaldi (1665) gave the same explanation of optics, Huygens is another scientist who gave an undulatory description of light. We also mention relevant technological achievements in the XVII century such as the invention of a microscope by the scientist Zacharie Janssen and the construction of the telescope by Galileo Galilei in 1610. This last and innovative invention of Galilei is at the origin of the discovery of sunspots which are nowadays used to study the phenomena of solar winds and geomagnetism generated by the sun on the ionosphere and the stratosphere. This same telescope is at the origin of the discovery by the same scientist of the satellites of Jupiter called Callisto Europa and Ganymede

To explain the two aspects of light, the wave aspect and the corpuscular aspect at the same time, or simply the Ondo corpuscular phenomenon of light, it is necessary to wait for the respective experiments of two eminent scientists fathers of the diffraction and the interference of the light. Francesco Maria Grimaldi (1618-1663) Italian physicist and *astronomer* introduces the term of *diffraction* for the first time to characterize *the* new phenomena he has just discovered, using a very narrow beam of light passing through a thin slit. The phenomenon of diffraction characterizes the wave aspect of light. After this discovery many scientists like Robert Hooke, Christian Huygens and Leonhard Euler studied the question. However, Isaac Newton, through numerous experiments on light, rejected the wave theory and successfully developed the corpuscular theory of light. This theory had several scientists who adopted it such as Pierre Simon Laplace and Jean Baptiste Biot. It is necessary to wait for the English scientist Thomas Young who studied in Göttingen the medicine and not the physics which, in 1800 presented a paper to the royal society arguing that the light is also a wave in movement and in 1803 it proved the phenomenon of the interferences of the light by splitting a single source by a screen on which it created two obturations (slits of dimension of the order of the wavelength).

From the 1950s, mathematical techniques, especially Fourier analysis, have allowed the introduction of new concepts such as image analysis, transfer functions, spatial filtering etc.. Advances in computer science have led to significant improvements in the calculation and realization of optical systems: aspherical lenses, polishing by ion bombardment, surface treatments (anti-reflection for example). Optical fibers have become essential in the field of telecommunications while thin films continue to be studied as light guides.

Optics has known a great advance thanks to the advent of LASER (Light Amplification by Stimulated Emission of Radiation). The first laser was built in 1960 and since then the range of available wavelengths has expanded considerably from infrared to ultraviolet. Lasers are now used everywhere: CDs, DVDs, steel cutting in industry, barcode reading, high-tech scalpel in surgery.

Thousands of optical systems for visualization are currently in use around the world: computer screens, watches, calculators, cell phones.

The holography (process of reconstruction of the wave front: recording of the phase and the amplitude of the wave diffracted by an object) which makes it possible to obtain splendid 3D images, is used nowadays in many applications: non-destructive analyses, data storage.

The most that optics has brought to fiber optic telecommunications would be the OOO (optical-optical-optical) switching process that uses the conversion of light into electrical signal

II. General introduction

Physical optics corresponds to a subject of physics that studies all the physical properties é of light and the phenomena that are specific to it, in the visible, infrared and ultraviolet domains. These phenomena are detectable either with the naked eye or through specific measurement techniques via appropriate measuring devices. The most well known phenomena in physical optics are light interference, diffraction, light polarization and light propagation through waveguides such as optical fibers and power lasers.

The appearance of the laser, since 1960, with its rather complex theory has given to physical optics a considerable contribution allowing to put in evidences the quoted properties with an exemplary ease. It is by using the laser that a vertiginous evolution has been obtained. With a spatial coherence and a temporal coherence by amplification followed by a reduction of the time to femto-second (10^{-15} s) that an enormous power was realized for the laser. This has allowed the development of 'interferometric techniques using high resolution spectroscopy such as photoluminescence spectroscopy (XPS); UV photon spectroscopy (UPS), photoluminescence (PL).

We also mention the progress made in integrated optics via optical fiber, in non-linear optics allowing to obtain wavelengths in the visible or even ultraviolet to from those in the infrared, and finally holography which is a photography technique allowing to reconstitute the relief of objects thanks to the interferences formed to record the properties of any object in the space with three dimensions.

Laser guidance is also used by the weapons industry in high-precision tanks and rockets. This technology has allowed the military to hit any point on the globe with precision e cision per square meter using laser guidance and satellite tracking. We are now witnessing laser-based air defense, covering a complete territory. Progress in this direction is still in its infancy.

Physical optics has also had a very important development with the advent of the Fourier transform which allows to pass from the vibration of light in time to optical spectra in wavelengths or frequencies. With these techniques, light has allowed researchers in materials physics to obtain extraordinary results by highlighting the properties of emission and abortion of light in particular by semiconductor materials for applications in electronics for the manufacture of electronic components such as diodes, transistors or microprocessors and optoelectronic components such as LEDs or solar photovoltaic cells.

For a decade now, light has become an unrivalled medium for the transmission of information. Its very high frequency has not only allowed the assurance of a speed of execution and a great quantity of information but also it has multiplied by 10^6 the number of subscribers in telephones and in Internet with, in addition, a very precious security. Thus were born and developed the optical communications by optical fibers by putting to the archives the communications via copper cables.

In this book we propose to study phenomena and physical properties of light after having studied geometrical optics in volume I which has just been published by the same publisher.

In the first chapter we will discuss and define the mathematical properties of waves in order to better understand the wave phenomenon of light. The optical wave will be written in the same way as an electromagnetic wave in electronics. What changes is only the numerical values of the frequencies or wave vectors.

In a second chapter we will study light interference, optical interferometry via multiple systems of generation of interference and interferometry. We will give examples of construction of fields of interference bangs and we will highlight, using a mathematical formalism, the difference of march, the interrange and the order of interference. Interferometric systems such as Michelson, Fabry-Perrot and Mach-Zehnder devices will be discussed in another volume.

The third chapter will be devoted to diffraction phenomena. We focus our attention on the Fraunhofer diffraction which will be easier to handle by introducing other techniques of light diffraction generation. The theory of diffraction will have been treated and enlivened by examples in the form of application exercises.

The fourth chapter will highlight optical polarization, which will provide insight into the geometric form by which the electric field representing the light wave propagates in two and three dimensional space. We describe linear polarization and circular polarization through elliptical polarization. We introduce the Jones formalism which will simplify the writing of a polarized wave, we will also describe the optical polarization by polarizers and wave blades. The Jones formalism will facilitate the understanding of such a phenomenon in multiple applications in optical microscopy for the study of minerals or in photoluminescence using a polarized laser to better define the sub-levels in III-V materials and others.

The fifth chapter will be devoted to Fresnel coefficients and factors. The Fresnel formulas (or Fresnel coefficients) describe the reflection and transmission of light when it falls on an interface between different optical media. In general, the Fresnel coefficients, defined on the interface between two media, are used in the evaluation of the optical behavior of optical multilayers, from the simplest of them, the blade with parallel face, to the most complex: anti-reflection layers, interference filters

In this book, the reader will find solved exercises after each end of chapter. This will allow a better understanding and fixation of the knowledge of the different courses.

We will give a general conclusion at the end of this work to list the most important results to be retained in this work and the future perspectives to be realized in optics.

Chapter I
Propagation of light waves

I. 1Introduction

The waves, correspond to the deformations of a material medium or to the displacements of excitations in this medium. A wave is generally produced by a localized shaking of a continuous medium; after its creation, the wave moves in the medium. This is what is commonly called wave propagation. Examples of these propagation phenomena are the waves at the surface of the sea and the undulations created on a stretched rope. But in optics called wave, we can not give the eye this phenomenon of propagation of light, we imagine it propagating in a straight line from one point to another. We see it reflected by meeting an obstacle, we note that it gives a phenomenon of diffraction if this obstacle is provided with an obturation and we also observe a phenomenon of interference if we split a source on an opaque screen. All these phenomena can be observed on a screen placed in front of the light beam.

I.2 Properties of light

I.2.1 Wave-corpuscular phenomenon of the light

The light can be considered as being the physical agent essential to the vision. It is also an electromagnetic wave with a double aspect: It is first a disturbance of space associated with the presence of an electromagnetic field that varies in space and time, the light is therefore part of electromagnetic waves. This property gives light an undulatory aspect. It is also a flow of particles called photons which gives light a second property known as the corpuscular aspect of light

I.2.2 Light sources

We can see the existence of two types of light sources. A primary source and a secondary source.

A source is defined as primary, if the body produces the light and emits it spontaneously as for example, the sun or a lighted candle.

A secondary source is defined as an object that does not produce light, but reflects the light it receives. It is said to be a diffusing body. This is the case of the moon lit by the sun.

I.2.3 Light production

Three types of light are produced by electrical excitations. The light produced is different spectrally from each other.

    a. Discharge lamp or spectral lamp:

Light is produced when a gas or a metallic vapor, under high or low pressure, is excited by an electric discharge. The atoms of the gas do not remain indefinitely in the excited states, but de-excite spontaneously by emitting light at frequencies characteristic of the gas contained in the lamp. We observe a discontinuous spectrum, called line spectrum (we obtain a band spectrum in the case of molecules). As an example, we can cite the mercury spectral lamp. It produces a light that approaches the blue-green. When this light passes through a dispersive medium, a grating for example, a discontinuous spectrum is obtained consisting of colored lines: yellow doublet at 0.5791 µm and 0.5770 µm, green at 0.5461 µm, blue at 0.4916 µm, indigo at 0.4358 µm and violet at 0.4047 µm. Fig. I.1

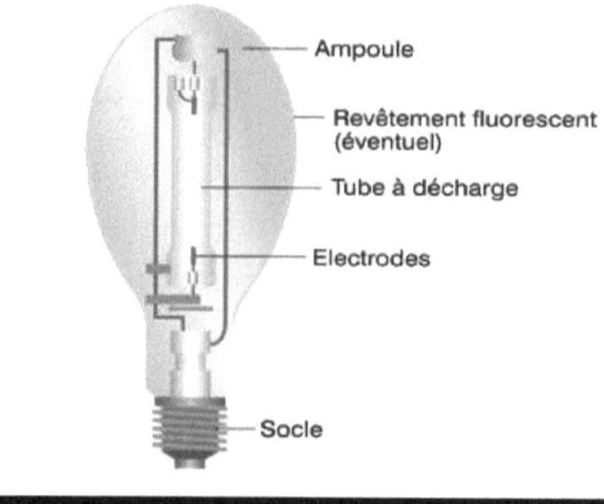

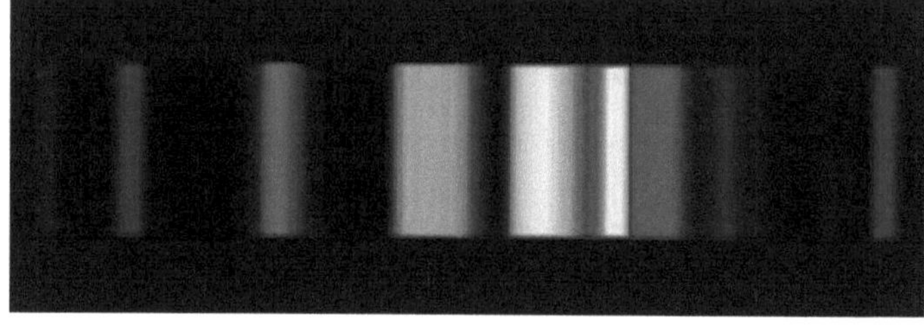

Fig. I.1 Discharge lamp emitting spectral light (top); Spectrum emitted by a discharge lamp (bottom)

b. Incandescent lamp : halogen lamp

A halogen incandescent lamp produces light in the same way as an incandescent lamp, by incandescing a tungsten filament, but in a small quartz glass bulb filled with halogen gases (iodine and bromine) at low pressure. This bulb works at high temperatures where the convection of the halogen gases allows the continuous regeneration of the filament, at least partially, which increases the life of the bulb

Light is emitted when a metal filament is heated to a high temperature. The spectrum in this case is continuous, it is consistent with the emission spectrum of a black body. This is the case of the conventional incandescent lamp that produces white light by bringing to incandescence a tungsten filament. Fig. I.2

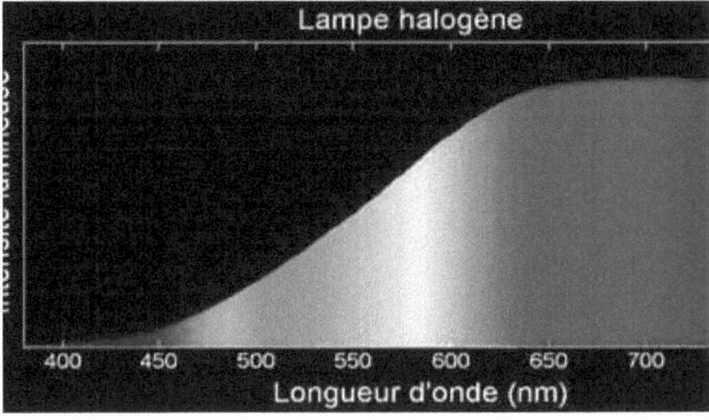

Fig.I.2 Incandescent lamp (top), continuous spectrum of a halogen lamp
(bottom)

c. Laser

The laser is the abbreviation of the words in English (Light
Amplification by Stimulated Emission of Radiation): it is an
optical cavity (called resonant cavity) in which a light amplifying
medium has been placed) fig. I.3. Four types of laser can be
mentioned: solid state lasers such as YAG (Yttrium Gallium
Garnet); gas lasers such as argon or He-Ne. The first is used in
medicine for retinal therapy, the second is used in the alignment of
cavities and in various practical work requiring a laser; the third
type are semiconductor lasers binary as laser GaAs or Ternary as
GaAsP or quaternary as InGaAsP. Semiconductor lasers are much

13

more used in fiber optic telecommunications. There are also chemical lasers, sources of radiation obtained by a chemical reaction, most often exothermic like the Iodine laser. Fig.I.3

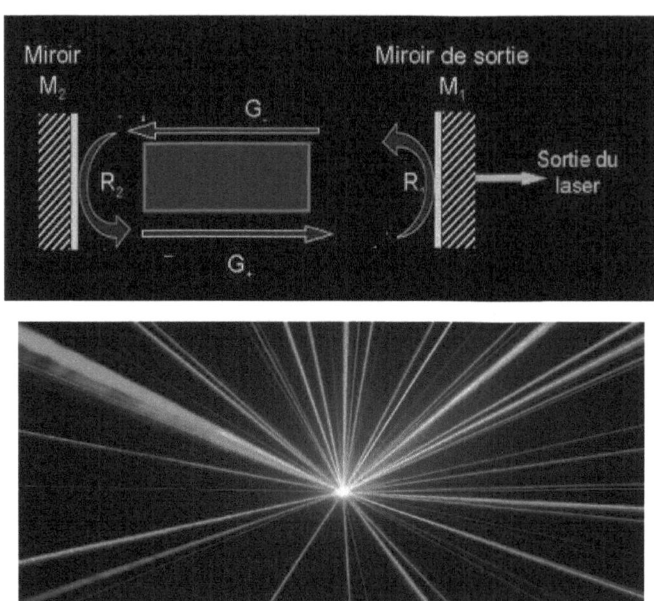

Fig. I.3 Schematic of a laser cavity (top), emitted laser light (bottom)

I.2 Mathematical formalism of a light wave

The wave is known by its four successive characteristics : its amplitude, which represents the "height" of the deformation with respect to the medium, its average position at the time of observation, its size around its average position, and its propagation speed or celerity. Fig. I.4. The accurate description of a wave requires its representation by a function which describes the deformation of the medium in which it propagates at each of its points. Thus if f (x, t) is the function that will represent the variable height of the deformation of a string, f(t, x) will be a function depending on space and time.

14

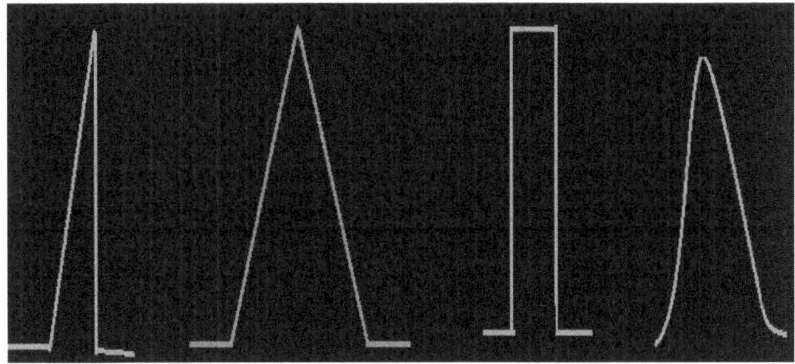

Fig. 1.4 - Schematic representations of transverse waves seen in profile.

1.3. Properties of an optical wave

Thus, as the optical wave evolves over time, it will be a function of time, represented by the symbol t. To locate the position of the wave in space, f (t, x) will also be a function of the coordinates **x**, **y**, **z**. If the movement is along a single direction called one-dimensional motion, a single coordinate, x for example, will be sufficient. For optics propagating in a vacuum, the speed would be the constant c. A parameter A_0 can represent the maximum height of the wave or its amplitude, this one can be an algebraic quantity, positive or negative, or even complex. We can then add the frequency, the pulsation and the period which will be essential parameters in the description of an optical wave and its propagation in the vacuum or in the media.

The phenomenon of optical wave propagation has properties that can be described by general considerations even before setting up and solving the equations of motion. The media in which the optical wave propagates are considered in principle to be non-viscous and non-dissipative media, not subject to external forces (in particular, the force of gravity is neglected). In these media, the wave propagates with a constant speed c and keeps an unchanged shape. Let us consider a transverse wave propagating along a direction x. Let us suppose that at the instant t_0 this wave has a shape around x_0, which will represent the abscissa of the maximum amplitude of the wave This wave is at any instant described by the function f (t, x), where t represents the instant of observation of the wave and x the observed abscissa. The value of f (t, x) gives the algebraic height (i.e. positive or negative) of the wave at the position x at time t in a vertical direction along the z axis. We can represent schematically the previous wave by the fig.1.2

At the time t_0 ; the function f (t_0 , x) gives the height of the wave for t_0 the value f (t_0 , x_0) represents at the time t_0 the height of the wave at the position x_0 , which corresponds in fact to its maximum height.

h_{max} = f (t_0 , x_0) (I.1)

And

$$h(t_0 , x_0) = f(t_0 , x) \quad (I.2)$$

At the following instants, this wave propagates along the x axis. Let us assume that it moves towards increasing x. Let us observe the wave at the time t_1 , such that $t_1 > t_0$. Fig.1.3 will represent the function f(t, x) at time t_1

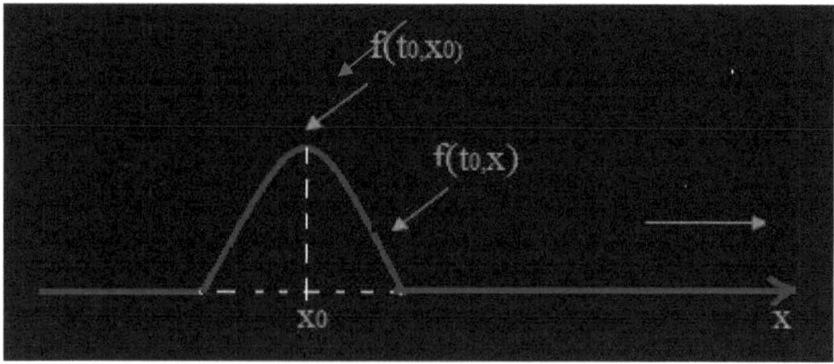

The wave in fig1.3 will keep its unaltered shape and is now localized around the abscissa x_1 , such that $x_1 > x_0$. Its shape is described by the function f (t_1 , x). Its maximum height corresponds to the value f (t_1 , x).$_1$

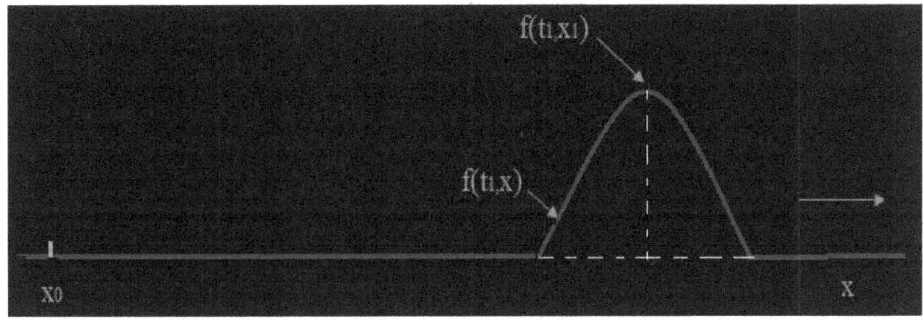

Figure 1.3 Representation of the wave at time t_1

By comparing the shape of the wave at t_0 and t_1 we can conclude that the wave is moving without modification of its shape; it is thus a global translation of the shape of the wave towards increasing x. In particular, if we compare the maxima at t_0 and t_1 , we should have :

$$f (t_1 , x_1) = f (t_0 , x_0). \tag{1.3}$$

The mathematical structure inherent in this equation can be highlighted as follows, let us introduce the time interval τ cut between the instants t_0 and t_1 that is:

$$\tau = t_1 - t_0. \tag{1.4}$$

The distance traveled by the wave during the time interval τ, is then

$x_1 - x_0 = c\tau.$ (1.5)

By eliminating t_1 and x_1 , Equation. (1.3) becomes:

$f(t_0 + \tau, x_0 + c\tau) = f(t_0, x_0).$
(1.6)

This means that in the previous' equation we can replace t_0 and x_0 by any other pair t and x :

$f(t + \tau, x + c\tau) = f(t, x).$
(1.7)

The shape of the wave, observed at time t, must be found at time t $+ \tau$ by an ensemble translation of all its constituent points by one distance.
The function f must depend only on the combination (x - ct) of t and x (or what is equivalent to: (t - x/c)):

$$f(t, x) = f(x - ct) \quad (1.8)$$

We can now check that if f has the previous structure then the' Eq. (1.7) is satisfied. Indeed, the first member becomes :

$f(t + \tau, x + c\tau)$
$= f((x + c\tau) - c(t + \tau)) = f(x - ct) = f(t, x) \quad (I.9)$
Which reproduces well the second member of the Equation. (1.7). These considerations can be repeated for waves propagating towards decreasing x. Figure I.4 and figure I.5 describe correctly this propagation towards decreasing x.

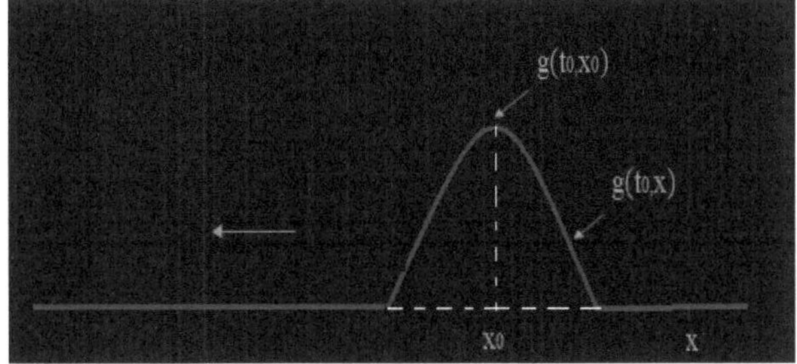

Fig. I.4 representation of the decreasing wave at position x0

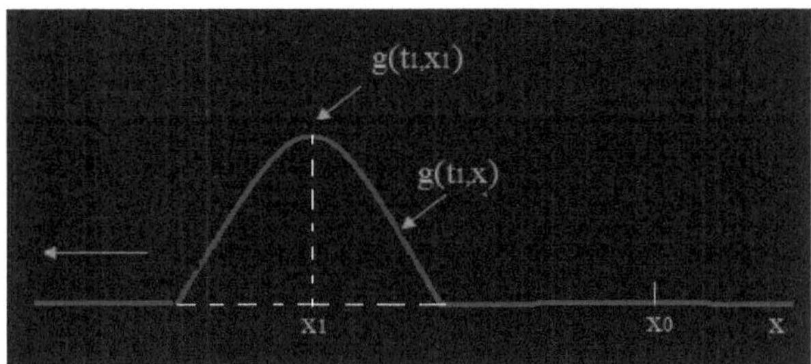

Fig. I.5 representation of the decreasing wave in position x1

I.4 Wave equation

The general expression can be put in the form :

$f(x, t) = f_1(t - \frac{x}{v})$ (I.10)

Let us check that the function s (x,t) given by ((I.8)) is a solution of the wave equation given by the equation

$\nabla^2 f = \frac{1}{v^2} f$ (I.11)

If ∇ *est* one-dimensional it will be the derivative with respect to the variable x evening $\nabla = \frac{d}{dx}$

Also:

$$-\frac{1}{v^2}\frac{df}{dt^2} = -\frac{1}{v^2}f$$

(I.12)

Thus we have:

$$\nabla^2 f - \frac{1}{v^2}\frac{df}{dt^2} = 0$$
(I.13)

Therefore f (x, t) is a solution of (I.13)

If we pose: u= t - $\frac{x}{v}$ then f_1(x, t) =f_1(u)

For this wave to keep its amplitude periodically constant we must have :

$$u_1 = u_2$$

That is: fig (I.5)

$$t_1 - \frac{x_1}{v} = t_2 - \frac{x_2}{v}$$
(I.14)

Equation (I.13) gives us :

$$x_2 - x_1 = v(t_2 - t_1)$$

We note that

$$v = \frac{dx}{dt} = \text{cte} \quad (I.15)$$

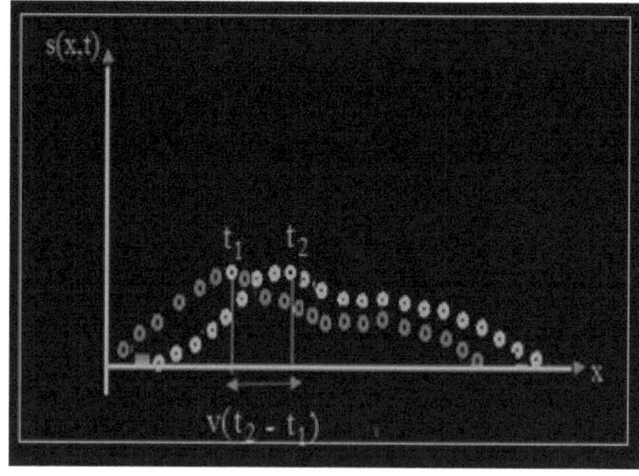

Fig. .I.5 : Propagation of a progressive wave depending on the two variables x and t

Thus v is the speed of propagation of the wave fig. I.5

We can have the regressive wave that is to say the one that is reflected by any obstacle and is written :

f (x, t)= $f_2(t + \frac{x}{v})$ it will also verify the wave equation :

$$\nabla^2 f_2 - \frac{1}{v^2}\frac{df_2}{dt^2} =$$
0
(I.16)

A plane optical wave is written in the form :
f (x, t)= $f_1(x,t)) = a \cos(\omega t -$
kx+φ)
(I.17)
a (amplitude), ω (pulsation). k (wave
vector). φ (phase à l'rigine) They are constants.
This equation is a solution of the differential equation :
$$\nabla^2 f - \frac{1}{v^2}\frac{df}{dt^2} =$$
0
(I.18)

$\nabla^2 f$ =$-k^2$f (x, t) and $-\frac{1}{v^2}\frac{df}{dt^2} = = \frac{\omega^2}{v^2}$f(x, t) (I.19)

Introducing the two derivation results in equation (I.17) we will have :

$$- k^2 \text{f (x,t)} = - \frac{\omega^2}{v^2}\text{f (x, t) (I.20)}$$

Equation (I.20) is called the dispersion equation, which relates the wave vector with the pulsation and the propagation velocity:
k =$\frac{\omega}{v}$
(I.21)
We can take the complex notation as follows:
s (x, t) = a $e^{-i(\omega t - kx + \varphi)}$
(I.22)

In fact it is the real part of s (x, t) which corresponds to the variations of f (x, t).

The period of the oscillation motion of the periodic function f (x, t) is of two kinds:

-Time period

$$T=\frac{2\pi}{\omega} \quad \text{(second) (I.23)}$$

-Spatial period also called wavelength :

$$\lambda=\frac{2\pi}{k} \quad \text{(meter) (I.24)}$$

The velocity v, wavelength λ and period T are related by the relation :

$$\lambda=v\,T \quad (I.25)$$

T is the temporal period and λ is the spatial period. Fig. I .6

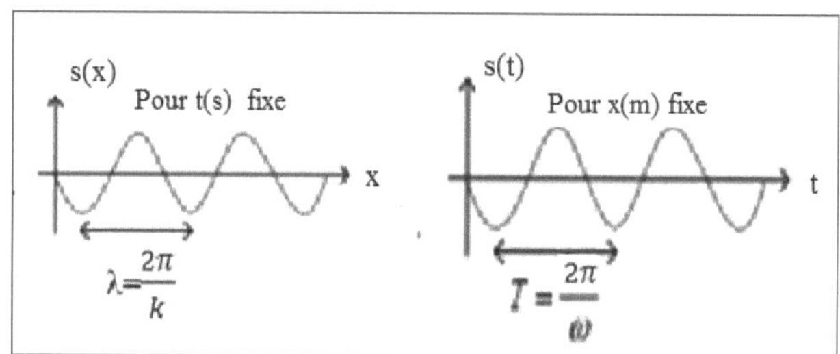

Fig. .I.6 Movement of the progressive wave in time and space

I.5. Propagation in space à three dimensions

S oit is a periodic plane wave f (r, t) whose general expression is written as :

$$f (r, t) =a \cos (\omega t - \vec{k}\vec{r}+\varphi)$$

(I.26)

Or its exponential form:

$$f (r, t) = ae^{-i(\omega t-\vec{k}\vec{r}+\varphi)} \qquad (I.27)$$

With f (r, t) = E (r, t) and or B (r, t) or E (r, t) and B (r, t)
represent respectively the electric field and the magnetic field

With the wave vector having three components in three
dimensional space.

$$\vec{k} = \begin{pmatrix} k_x \\ k_y \\ k_z \end{pmatrix} \quad \text{and} \quad \vec{r} =$$

$$\begin{pmatrix} x \\ y \\ z \end{pmatrix} \qquad\qquad (I.28)$$

Thus their scalar product becomes :

$$\vec{k}\vec{r} = k_x\, x + k_y y + k_z z$$
(I.29)

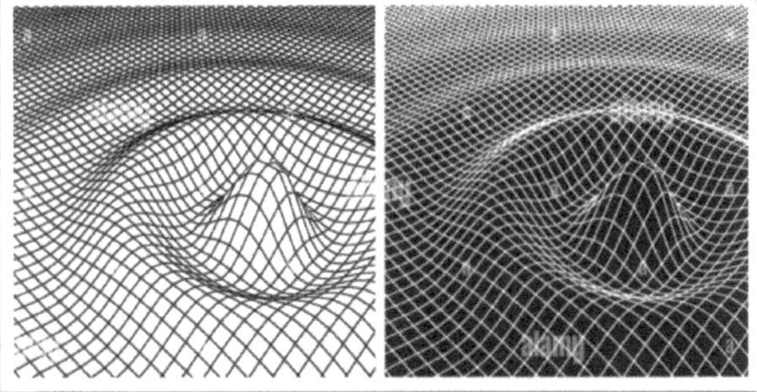

I.6 Applications to Chapter I

Exercise I.4.1

We ask to check which expressions are waves by verifying the wave equation in the following:

1) $g_1(z, t) = 5Ce^{-2(z-vt)}$;
2) $g_2(z, t) = 2 \sin 4(z - vt)$;
3) $g_3(z, t) = \dfrac{8}{a(z-vt)^2 + 5}$;
4) $g_4(z, t) = Ke^{-3(z^2+vt)}$;
5) $g_5(z, t) = K \cos(2z) \sin(3vt)$;
6) $g_6(z, t) = K \cos(3(z-vt))$.

Solution Exercise I.4.1

If $g(z, t)$ is a wave function it must verify the following wave equation:

$$\nabla^2 g(z,t) - \frac{1}{v^2}\frac{dg(z,t)}{dt^2} = 0$$

In this case we have $g(z, t)$ which depends only on the variable z, so :

$$\nabla^2 = \frac{\partial^2}{\partial z^2}$$

1. for the function $g_1(z, t) = 5Ce^{-2(z-vt)}$ we have :

$$\frac{\partial^2 g_1(z,t)}{\partial z^2} = -10Ce^{-2(z-ct)}$$

$$\frac{\partial^2 g_1(z,t)}{\partial t^2} = -10C\, v^2 e^{-2(z-vt)}$$

So, using the previous results we will have :

$$\frac{\partial^2 g_1(z,t)}{\partial z^2} - \frac{1}{v^2}\frac{\partial^2 g_1(z,t)}{\partial t^2} = 0$$

So $g_1(z, t) = 5Ce^{-2(z-vt)}$, verifies a wave equation.

2. For the function $g_2(z, t) = 2 \sin(4(z-vt))$, we have :

$$\frac{\partial^2 g_2(z,t)}{\partial z^2} = 8\sin(4(z - vt)),$$

$$\frac{\partial^2 g_2(z,t)}{\partial t^2} = 8v^2\sin(z\text{-}vt)$$

Thus we will have :

$$\frac{\partial^2 g_2(z,t)}{\partial z^2} - \frac{1}{v^2}\frac{\partial^2 g_2(z,t)}{\partial t^2}=0$$

Similarly this function(z, t) = 2 sin(4(z-vt)), also verifies the wave equation.

3. For the function $g_3(z, t) =\dfrac{8}{a(z-vt)^2+5}$ we have :

$$\frac{\partial^2 g_3(z,t)}{\partial z^2} = \frac{-32a^2}{(a(z - vt)^2 + 5)^2}$$

$$\frac{\partial^2 g_3(z,t)}{\partial z^2} = \frac{-32a^2v^2}{(a(z - vt)^2 + 5)^2}$$

We can see that this function $g_3(z, t) =\dfrac{8}{a(z-vt)^2+5}$also verifies the conditions of a wave function:

$$\frac{\partial^2 g_3(z,t)}{\partial z^2} - \frac{1}{v^2}\frac{\partial^2 g_3(z,t)}{\partial t^2}= 0$$

4. For the function $g_4(z, t) =Ke^{-3(z^2+vt)}$ we have :

$$\frac{\partial^2 g_4(z,t)}{\partial z^2} =\text{-}12z^2 e^{-3(z^2+vt)}$$

$$\frac{\partial^2 g_4(z,t)}{\partial t^2} = -9v^2 e^{-3(z^2+vt)}$$

We can see that this function $g_4(z, t) =Ke^{-3(z^2+vt)}$ does not verify the conditions of a wave function.

5. For the function $g_5(z, t) =K \cos (2\ z) \sin (3\ v\ t)$,we have :

$$\frac{\partial^2 g_5(z,t)}{\partial z^2}= 4K\cos (2z) \sin (3v\ t)$$

$$\frac{\partial^2 g_5(z,t)}{\partial t^2} =9v^2K\cos (2z) \sin (3v\ t)$$

The function $g_5(z, t) = K \cos (2\,z) \sin (3\,v\,t)$ does not satisfy the conditions of a wave function either.

6. For the function $g_6(z, t) = K \cos (3(z\text{-}vt))$, we have

$$\frac{\partial^2 g_6(z,t)}{\partial z^2} = 9K \cos (3(z\text{-}vt))$$

$$\frac{\partial^2 g_5(z,t)}{\partial t^2} = 9Kv^2 \cos (3(z\text{-}vt))$$

La fonction $g_6(z, t) = K \cos (3(z\text{-}vt))$ verifies well the conditions of a wave function.

Exercise I.4.2

Let a wave whose equation is written :

$$z(t, x) = 0.08\left[\frac{\pi}{4}(8x - 20t) - \frac{\pi}{6}\right]$$

With z and x in sets and time t in seconds. Find

a) The wavelength, frequency and speed of propagation of the wave.
b) The speed and acceleration of a particle located in the path of the wave has x= 0.5m and t= 0.01 s.

Solution exercise I.4.2

By development of the wave equation we will have :

$$z(x, t) = 0.08\left[\frac{\pi}{4}(8x - 20t) - \frac{\pi}{6}\right] = 0.08\left[2\pi x - 5\pi t - \frac{\pi}{6}\right]$$

$$z(x, t) = 0.08\left[2\pi x - 5\pi t - \frac{\pi}{6}\right]$$

By comparing this last equation with the canonical form of a plane wave which is written :

$$z(x, t)=A[kx - \omega t + \varphi]$$

We will have:

$$k=2\pi \text{ , and } \omega = 5\pi \text{ and } \varphi=-\frac{\pi}{6}$$

Exercise I.4.3

1. A wave motion is 6m in 20s.
Knowing that 2 vertices of the wave are separated by 40 cm,
calculate the frequency of the wave.
Write the equation of a point M located at 1.8m from the source.

    a. The velocity of the wave is calculated as follows:

$$v=\frac{x}{t}=\frac{6}{20}=0.3 \text{ m/s}$$

    b. Two vertices separated by 40 cm means the wavelength

$$l = 0.4m$$

    c. The frequency of the wave is :

$$f=\frac{v}{\lambda}=\frac{0.3}{0.4}=0.75 \text{ Hz}$$

With the pulsation $\omega = 2\pi f = 2 * 3.14 * 0.75 = 4.71$ rd/s

A point located at 1.8 m from the source has a phase shift of

$$\varphi = \frac{2\pi x}{\lambda}=\frac{2*3.14*1.8}{0.4}=$$

    d. Equation of the wave

$$y=y_0 \sin(4.71 \text{ t}+28.86)$$

Exercise I.4.4

Let the transverse wave described in the figure opposite. Its
propagation speed is 40 cm/s towards the right.

Determine:

(a) frequency;

b) the phase difference in radians between points 2.5 cm apart;

c) the time required for the phase at a given point to change by60°
;
d) the velocity of a particle at the point P at the represented time

Solution
The transverse wave is sinusoidal in shape and propagates at $v =$ 40 cm/s to the right. The analysis of the figure allows to obtain the other necessary data.

a) As $l = 4$ cm, we obtain
$$l = \frac{v}{f}f = \frac{v}{\lambda}$$
A.N v=40 cm/s=0.4m/s and $l = 4$ cm =0.04m

So $f = \frac{0.4}{0.04} = 10$ Hz

b) at any position x of the wave, the phase becomes :
$$\varphi = \frac{2\pi x}{\lambda}$$
If the distance between two points is $\Delta x = 2.5$ cm, the phase variation between these two points is given by :

$$\Delta\varphi = \frac{2\pi\Delta x}{\lambda} = \frac{2\times3.14\times2.510^{-3}}{0.04} = 0.3925\text{rd}$$

c) The time required for the phase to change at a point from
$$\Delta\varphi = 60° = \frac{\pi}{3}\text{ rd}$$
$$\Delta\varphi = \frac{2\pi\Delta x}{\lambda} = \frac{2\pi v\Delta t}{\lambda} = \frac{\pi}{3}$$
This gives:
$$\Delta t = \frac{\lambda}{6v} = \frac{0.04}{6\times0.4} = 16.66\text{ms}$$

Chapter II
Diffraction of light

Diffraction of light

II.1 Theory of diffraction

II.1.1 Introduction

The diffraction was obtained by an electron beam or by any particle. It is well characterized by electromagnetic waves (thus light waves) as well as by mechanical waves (sound waves, waves on the surface of water ...). Obtaining diffraction by a light wave contradicts the laws of geometric optics for which the propagation of light is straight in a homogeneous and transparent medium. Diffraction is a physical phenomenon that occurs when a light wave propagating in a given direction meets an obstacle with a slit whose dimensions are of the order of the wavelength, but the effects of this obstacle are manifested and will be observable only when the dimensions of this obstacle are of the order of magnitude of the wavelength. The diffraction of light explains the optical phenomenon related to the wave properties of light, so the effects of diffraction can not be explained by geometric optics. From a historical point of view, diffraction was discovered with light in 1665 by Grimaldi. The phenomenon of light diffraction was known, but Grimaldi was the first to observe it (by looking at a rainbow) and describe it under experimental conditions. Grimaldi's work on optics is contained in his treatise Physico-mathesis de *lumine, coloribus et iride*, whose title can be translated as : *Physical knowledge on light,* colors and the rainbow. The treatise was published in 1665 in Bologna, after his death. Diffraction was correctly interpreted as a wave behavior by Huygens, then studied by Fresnel and Fraunhofer following Young's experiments (Young's slits).

II.1.2 Diffraction intensity

Consider a slit in the plane O x y, of width **a** in the x direction, and infinite along y. This slit is illuminated by an incident plane wave propagating in the direction Oy. We place ourselves in the simple case where the incident rays from the light source are parallel to the axis of propagation Oy. To simplify the calculation of the amplitude and intensity of the light diffracted by the slit of width a, we place ourselves in the case of an observation of diffracted waves at infinity ie in the case of a Fraunhofer diffraction or far field diffraction. (Fig.II 1)

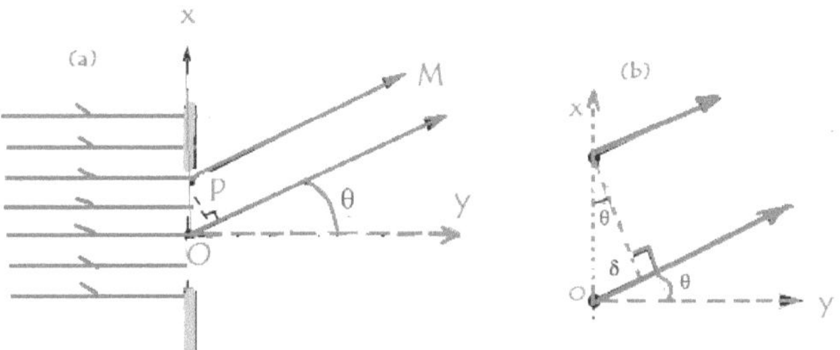

Fig. II.1 Diffraction of light rays through a thin slit

Consider Fig. II.1.b which is an enlargement of the slit in Fig. II.1a. We note that the step difference δ between the wave diffracted in the θ direction at point P and that diffracted in the same direction at point O is given by the expression :

$\delta = - x \sin \theta$ (II.1)

The phase difference φ is given by the expression :

$\varphi = \dfrac{2\pi}{\lambda} x \sin \theta$ (II.2)

We can then express the amplitude of the wave for a given position x by the relation :

$A(x) = A_0 \exp(-j\varphi) = A_0 \exp\left(-j\dfrac{2\pi}{\lambda} x \sin \theta\right)$ (II.3)

The total amplitude of the diffracted wave will be obtained by the integral over the entire extent of the aperture or :

$$A(M) = A_0 \int_{-\frac{a}{2}}^{\frac{a}{2}} \exp\left(-j\frac{2\pi}{\lambda} x \sin\theta\right) dx \quad \text{(II.4)}$$

$$A(M) = A_0 \frac{\lambda}{2\pi j \sin\theta} \left| \begin{array}{c} \left(\exp\left(j\frac{2\pi}{\lambda}\frac{a}{2}\sin\theta\right)\right. \\ \left. - \exp\left(-j\frac{2\pi}{\lambda}\frac{a}{2}\sin\theta\right)\right) \end{array} \right| \quad \text{(II.5)}$$

$$A(M) = A_0 \frac{\lambda}{2\pi j \sin\theta} \left| \begin{array}{c} \left(\exp\left(j\frac{\pi a}{\lambda}\sin\theta\right)\right) \\ - \exp\left(-j\frac{\pi a}{\lambda}\sin\theta\right)) \end{array} \right| \quad \text{(II.6)}$$

The expression (II.6) can be put in the following form:

$$A(M) = A_0 \frac{\sin\left(\frac{\pi a}{\lambda}\sin\theta\right)}{\frac{\pi a}{\lambda}\sin\theta} \quad \text{(II.7)}$$

The intensity of the light wave at point M will be obtained by raising the expression (II.7) to the square, i.e. :

$$I = I_{M0} \frac{\sin^2\left(\frac{\pi a}{\lambda}\sin\theta\right)}{\left(\frac{\pi a}{\lambda}\sin\theta\right)^2} \quad \text{(II.8)}$$

II.1.3 Discussion of the diffraction intensity

Thus according to the expression (II.8) the minimums of the diffraction intensity are obtained when the numerator cancels out, i.e. :

$$\sin^2\left(\frac{\pi a}{\lambda}\sin\theta\right) = 0 \quad \text{(II.9)}$$

This gives

$$\frac{\pi a}{\lambda} \sin\theta = m\pi \qquad\qquad \text{(II.10)}$$

Thus we obtain :

$$\sin\theta = m\frac{\lambda}{a} \qquad\qquad \text{(II.11)}$$

On the other hand, the intensity maxima will be obtained by deriving the expression (II.8)

We then put $x = \frac{\pi a}{\lambda} \sin\theta$ and the expression finally becomes :

$$I(x) = I_0 \frac{\sin^2 x}{x^2} \qquad\qquad \text{(II.12)}$$

Therefore, the derivative of this function gives :

$$\frac{dI(x)}{dx} = \frac{\sin x}{x}\left(\frac{x\sin x - \cos x}{x^2}\right) \qquad\qquad \text{(II.13)}$$

So if we take $\frac{dI(x)}{dx} = 0$ we will have

$$\frac{x\cos x - \sin x}{x^2} = 0$$
(II.14)

Or :

$$x \cos x - \sin x = 0$$

Which finally gives :

$$Tang\ x = x$$
(II.15)

This equation can have a graphical solution by the simultaneous plots e s of two functions $y_1 = tang\ x$ and $y_2 = x$. The intersection points of these two graphs are solutions of the equation The intersection points of y_1 and y_2 are obtained when

$$x= (m + \frac{1}{2})\pi$$

Figure II.2 gives a graphical representation of the function (II.15)

We then obtain the heights of the secondary maximums I_S relative to the intensity of the central maximum I_0 as follows:

$$I_S = \frac{I_0}{(m+\frac{1}{2})^2\pi^2} \tag{II.16}$$

The following table gives the different values calculated for the first five secondary maximums

Thus the first secondary maximum gives :

$$\frac{I_S}{I_M} = \frac{1}{(1+\frac{1}{2})^2\pi^2} = 0.0472$$

This value is equal to the error value obtained by experiment

This intensity decreases with the increase of $\frac{\pi a L}{\lambda D}$ because there are less and less sources that are in constructive interference to these differences of march.

We can calculate the different secondary maxima of the diffraction as follows:

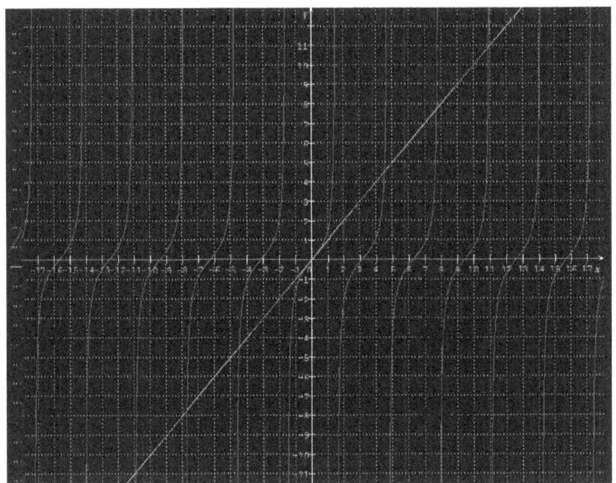

Fig. II.2 Graphical representations of the solution of equation (II.14)

$$\frac{I_S}{I_M} = \frac{1}{(m+\frac{1}{2})^2\pi^2} = \begin{cases} I_{s1} = 0.0471I_0 \\ I_{s2} = 0.0171I_0 \\ I_{s2} = 0.008I_0 \\ I_{s2} = 0.005I_0 \\ I_{s2} = 0.003I_0 \end{cases}$$

II.1.3 Graphical representation as a function of $sin\theta$

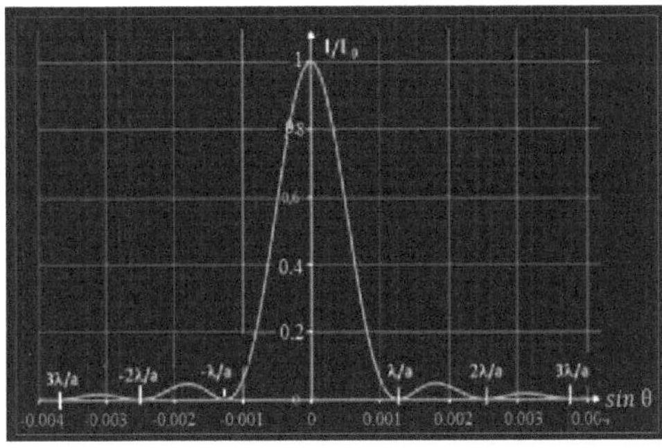

Fig. II.3 Diffraction pattern showing the relative intensity I/I_0 as a function of $sin\theta$

By observing the experimental set-up fig. II.4 we can put the expression (II.8) in the following form :

$$I = I_{M0} \frac{sin^2(\frac{\pi a L}{\lambda D})}{(\frac{\pi a L}{\lambda D})^2} \qquad (II.17)$$

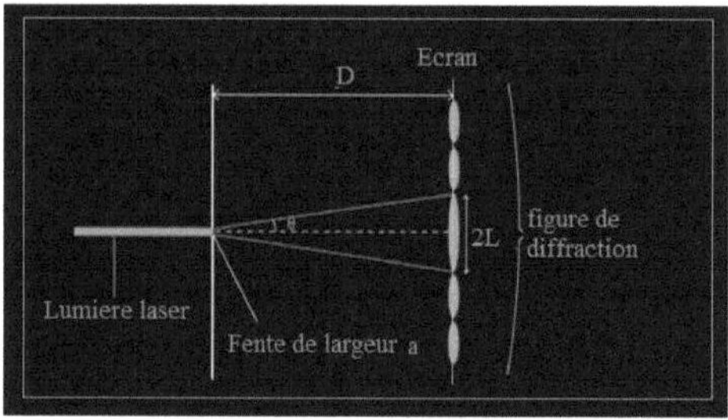

Fig. II.4 graphic representation of the intensity of an optical wave diffracted by a rectangular slit

The expression (II.17) contains all the parameters that can vary the diffraction pattern:

The minima of the diffraction pattern will be obtained when the numerator cancels, i.e. :

$$sin^2(\frac{\pi a L}{\lambda D}) = 0 \qquad (II.18)$$

Or :

$$\frac{\pi a L}{\lambda D} = m\pi \qquad (II.19)$$

Or again

$$\frac{L}{D} = m\frac{\lambda}{a} \qquad (II.20)$$

With :

1. λ the wavelength

2. a the width of the slot :

3. e D The slit-screen distance

II.2 4 Graphical representation of the intensity of a diffracted light wave as a function of $\varphi=\frac{\pi aL}{\lambda D}$

We consider a slit which is an opening of width a and infinite length, centered on the origin (the slit extends from $-a/2$ to $a/2$ in the *x-axis*). One of the approximations considered here is that we place ourselves in the case where the screen is located at infinity (Fraunhofer diffraction), that is to say that the rays arriving at any point of the screen are considered as parallel. This is the case if the screen is placed a few meters from the slit Figure II.5 gives a graphical representation of the diffraction intensity for $I_0 = 5$, $\lambda=0.5\mu m$ and a $=1.6$ mm.

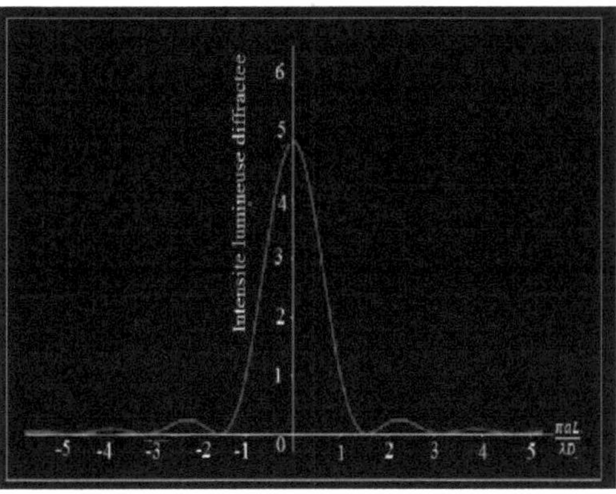

Fig. II.5 graphic representation of the intensity of the light wave diffracted by a thin slit.

II.3. Diffraction through a circular aperture

The calculation of the intensity diffracted in the direction □ □□ donnée by a circular slit of radius R will be different from that of a thin rectangular slit. We use the Bessel functions to get around this. The intensity diffracted at infinity by a circular aperture illuminated under normal incidence by a beam of monochromatic light can be determined as follows: The problem is rotationally symmetric about the axis Oz of the diaphragm of radius

R. The diffracted pattern will have the same symmetry. Let us therefore consider a direction characterized by the angle θ that a diffracted ray makes with the axis Oz, corresponding to the direction of incidence of the light ray. Fig. II.6

Or :

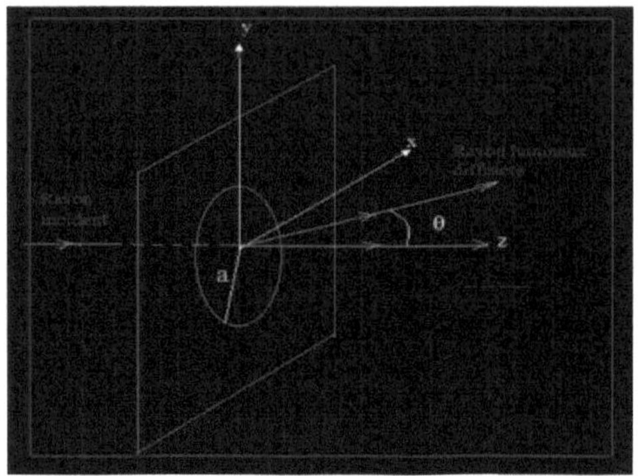

Fig. II.6 Diagram of the diffraction of light by a circular slit

In such a problem and to arrive at the end of this type of diffraction, we must use the cylindrical coordinates We will have :

$$\vec{u}=\cos\theta\vec{e}_z + \sin\theta\vec{e}_x$$

Thus we can write the amplitude of the diffracted vibration in the direction of the vector $\vec{u}$ for a surface element dS as follows:

$$dA=A_0 \; dSe^{j\varphi} \tag{II.21}$$

In cylindrical coordinates the surface element dS is written :

$$dS=r \; dr \; d\Phi \tag{II.22}$$

To integrate, we introduce the limits of the variables r and Φ are :

$$0\leq r \leq a$$

$$0 \leq \Phi \leq 2\pi \tag{II.23}$$

The relation giving the phase φ as a function of the step difference δ and the wavelength λ is written as follows:

$$\varphi = \frac{2\pi\delta}{\lambda} \tag{II.24}$$

We also have the difference of march which will be expressed as follows:

$$\delta = \overrightarrow{OM}\,\vec{u} \tag{II.25}$$

And as

$$\overrightarrow{OM} = cos\Phi\,\vec{e}_x + sin\Phi\,\vec{e}_y \tag{II.26}$$

The scalar product of the relation (II.25):

$$\delta = rcos\Phi sin\theta \tag{II.27}$$

$$A = \int_0^{2\pi} \int_0^R A_0 e^{j\frac{2\pi}{\lambda}rcos\Phi sin\theta} rd\theta d\Phi \tag{II.28}$$

The light intensity of the diffraction of the circular slit is given by :

$$I = AA^* \tag{II.29}$$

$$I(\theta) = I_{max} \left| \int_0^d \rho d\rho \int_0^{2\pi} e^{j\frac{2\pi}{\lambda}\rho cos\Phi sin\theta} d\Phi \right|^2 \tag{II.30}$$

II.3.1 Introduction of   the Bessel function

We define the Bessel function of order zero as follows:

$$2\pi J_0(\frac{2\pi}{\lambda}\rho sin\theta) = \int_0^{2\pi} e^{j\frac{2\pi}{\lambda}\rho cos\Phi sin\theta} d\Phi \tag{II.31}$$

The calculation is done by introducing the first Bessel function $J_1(x)$

J_0 is the Bessel function of order zero, it is linked e to the Bessel function of order 1 by the relation:

$$xJ_0(x)=\frac{d(xJ_1(x))}{dx} \qquad (\text{II.32})$$

Hence the diffraction varies as follows

$$I(\theta) =I_{max}(\frac{2J_1(\frac{2\pi}{\lambda}asin\theta)}{\frac{2\pi}{\lambda}asin\theta})^2=I_{max}(\frac{2J_1(m)}{m})^2 \qquad (\text{II.33})$$

Posing:

$$m=\frac{2\pi}{\lambda}asin\theta \qquad (\text{II.34})$$

II.3.2   Value of the diffraction intensities of a circular slit

We evaluate the variation of $I(\theta)$ taking into account the calculated values of the Bessel functions, the table (II.1) summarizes these values.and the fig.II.7 gives a vision of what observable on a screen when one uses a circular slit.

Table (II.1) interference rings obtained by a circular slit

$sin\theta$	$\dfrac{I}{I_0}$	Interfringe type
$1.22\dfrac{\lambda}{a}$	0	1er dark ring

$1.63\dfrac{\lambda}{a}$	0.017	1^{cr} shiny ring
$2.23\dfrac{\lambda}{a}$	0	2i^{eme} dark ring
$2.68\dfrac{\lambda}{a}$	0.04	2i^{eme} shiny ring
$3.23\dfrac{\lambda}{a}$	0	3i^{eme} dark ring
$3.70\dfrac{\lambda}{a}$	0.0016	3er shiny ring

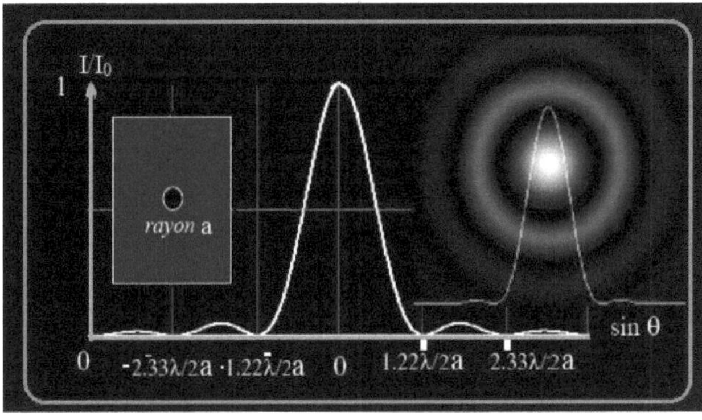

Fig.II.7 : curve and rings of interference given by a circular slit

II.3 Applications to diffraction
II.3.1 Fine slit
II.3.1 Exercise

An experiment is performed using a laser, a slit of adjustable width and a white screen. The device is represented on fig. II.8

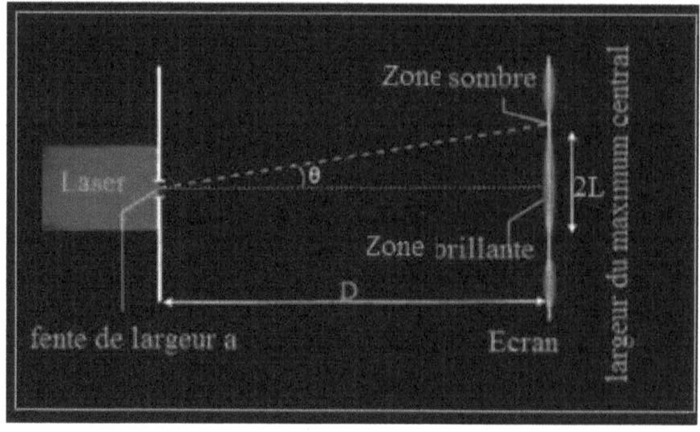

Fig. II.8 Representation of the experimental device of diffraction by a slit

1. what is the name of the observed phenomenon?
2. Calculate the angleθ in radian considering this angle very small.
We give
a = (0.20 ± 0.01) mm; D = (2.00 ± 0.01) m; 2L = (12.6 ± 0.1) mm

3. a Recall the relationship between the wavelengthλ and the angle θ
3. b Deduce the wavelengthλ .
3. c. The uncertainty of the wavelength measurementλ is evaluated by Δ(λ), Calculate the uncertainty Δ(λ) on the wavelength of the laser.

 4. Recall the relation betweenλ , c (celerity of the light in the vacuum) andν (frequency of the light radiation) ?
 Indicate their units in the international system.

5. Indicate how the width l varies when we
5.a: replaced the laser emitting red light with a laser emitting blue light ?
5.b: decreased the width of the slit a ?

Answer to question 1.1

If the light passes through a thin slit, the phenomenon of diffraction is observed.

Answer to question 1.2

The. Trigonometric relation in fig. II.4 is written as follows

$$\tan = \theta \approx \theta \frac{L}{D} = \frac{6.3 \, 10^{-3}}{2} . = 3.155 \, 10^{-3} \text{ rad}$$

Answer to question 1 3.a

We want to calculate the wavelength of the diffracted light: The importance of the phenomenon of diffraction is related to the ratio of the wavelength to the dimensions of the aperture (or the obstacle).
We have:

$$\text{Tang}\theta \approx \theta = \frac{L}{D} = \frac{\lambda}{a}$$

Answer to question 1 3.b

This gives

$$\lambda = a \frac{L}{D} =$$

$$\lambda = a\theta$$

$$\lambda = 0.200 \, 10^{-3} \times 3.155 \, 10^{-3} = 0.631 \, 10^{-6} \text{m} = 0.631 \; \mu m$$

This is the wavelength of a He-Ne laser in the red.

Answer to question1 3.c

Calculation of uncertainty on the wavelength λ

According to the relation giving the expression of λ , we will have :

$$\frac{\Delta\lambda}{\lambda}=\frac{\Delta a}{a}+\frac{\Delta L}{L}+\frac{\Delta D}{D}$$

$$\frac{\Delta\lambda}{\lambda}=\frac{0.01}{0.20}+\frac{0.1}{12.6}+\frac{0.01}{2}=0.0179$$

$$\Delta\lambda=0.0629\times0.631 10^{-6} = 0.039\ 10^{-6} m$$

Thus

$$0.6310 -0.039 < \lambda < 0.6310 +0.039\ (\mu m)$$

$$592 < \lambda < 670\ (nm)$$

Answer to question1 4

Let's determine the required relationships

We know from the relationships established in Chapter I that :

$$\lambda = c.T$$

$$\lambda = \frac{c}{f}$$

With c the speed of light in vacuum

$$c=3.10\ m/s^8$$

λ : the wavelength in meters

And f the frequency in Hz

Answer to question 5

The blue and red radiations in the vacuum are approximately

$$\lambda_{bleu} \approx 400\ nm$$

And

$$\lambda_{rouge} \approx 800 \text{ nm.}$$

- If we replace the red light emitting laser with a blue light emitting laser?

a. If we switch from blue to red light by switching to green light, the width of the central maximum will increase (see fig. II.9)

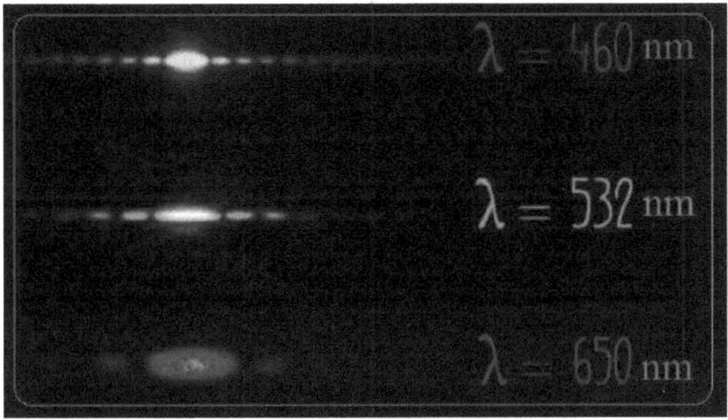

Fig. II.9 Enlargement of the central maximum when the color changes from blue to green and red respectively

b. If we decrease the width of the slit the figure widens and the central maximum spreads. The width of the central maximum is inversely proportional to the width **a** of the slit.

II.3.2 Exercise

We place a hair on the path of a beam of light of wavelength 638nm and we observe a diffraction phenomenon on a screen placed at 180cm from the hair.
We then see that the central spot has a width of 9.00cm.

1. Determine the angular deviation θ associated with this diffraction in radian. We can consider that $\tan(\theta)=)=\theta$

2. Deduce the thickness of the hair.

$$\text{Tang}= \theta = \frac{L}{D} = \frac{4.5\,10^{-2}}{180\,10^{-2}} = 0.025 \text{ rad}$$

On the other hand, we have :

$$\text{Tang}= \theta = \frac{\lambda}{a} = 0.025$$

We deduce that the thickness of the hair noted a will be:

$$a = \frac{\lambda}{\theta} = \frac{638\,10^{-9}}{0.025} = 25.52\ 10^{-6} = 25.52\mu m$$

II.3.3 Exercise

To determine the diameter **a** of an optical fiber, a laser beam of wavelength $\lambda = 650$nm is passed through it. A CCD sensor is placed at a distance $D = 2.0$ m from the exit of this optical fiber and the record below is obtained (the x-axis is graduated in cm).Fig. II ;10

1. What does the variation of the intensity of the curve represent
2. Determine the ratio of the intensities of the central peak with the first secondary. Can we know it theoretically? explain.
3. What is the value of the angular deviation θ in this figure?
4. Determine the thickness of this optical fiber

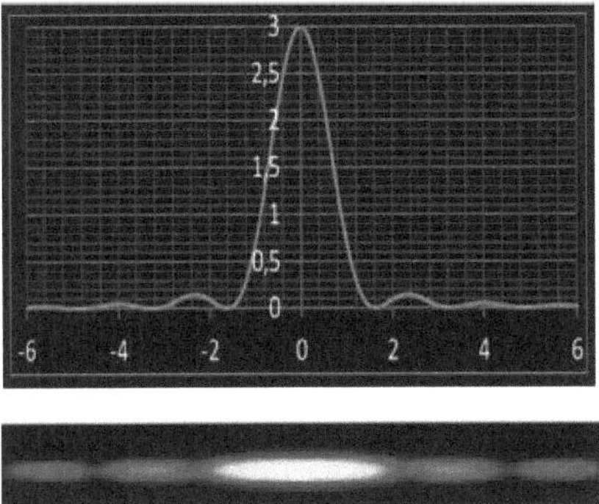

Fig. II.10 Diffraction pattern recorded by the CCD camera and corresponding diffraction image

Solution of the exercise II.3.3.3

Answer to question II.3.3.1

The intensity variation represents a diffraction pattern with a central maximum of width 3.6 cm and two secondary maxima symmetrical with respect to the O y axis representing the intensity of the light diffraction.

Answer to question II.3.3.2

The ratio of the intensities

$$R= \frac{I_S}{I_M} =\frac{0.141}{3}=0.47$$

This can indeed be deduced theoretically by calculation as follows:

First, we have the diffraction intensity of the light which is written :

$$I =I_{M0} \; \frac{sin^2 (\frac{\pi a L}{\lambda D})}{(\frac{\pi a L}{\lambda D})^2}$$

We have seen in the course of the diffraction that the heights of the secondary maxima are such that

$$\frac{I_S}{I_M} = \frac{1}{(m+\frac{1}{2})^2\pi^2}$$

Thus the first secondary maximum gives :

$$\frac{I_S}{I_M} = \frac{1}{(1+\frac{1}{2})^2\pi^2} = 0.0472$$

This value is equal to the error value obtained by experiment

This intensity decreases with the increase of $\frac{\pi a L}{\lambda D}$ because there are fewer and fewer sources that interfere constructively with these differences in rate.

We can calculate the different secondary maxima of the diffraction as follows:

$$\frac{I_S}{I_M} = \frac{1}{(m+\frac{1}{2})^2\pi^2} = \begin{cases} I_{s1} = 0.0471 I_0 \\ I_{s2} = 0.0171 I_0 \\ I_{s2} = 0.008 I_0 \\ I_{s2} = 0.005 I_0 \\ I_{s2} = 0.003 I_0 \end{cases}$$

II.3.2 Circular Slot Exercises
Exercise II.3.4

We consider a box with a circular hole of radius a and illuminated by a monochromatic source placed at the object focus of a lens of focal length f.

1. What is the radius of the light spot at the bottom of the box without diffraction?

2. What is the radius of the spot if there is light diffraction at the bottom of the box?

Since the light rays remain parallel with the axis of the hole section then we have

$$\text{Light spot} = \text{Hole section} = a$$

Considering the diagram of the experimental device of the diffraction with distance hole bottom equal to the focal length of the lens f :

$$\tan \theta = \frac{R_{tache\ lumineuse}}{f} = 1.22\frac{\lambda}{a}$$

$$R_{tache\ lumineuse} = 1.22\frac{\lambda f}{a}$$

Exercise II.3.5

What is the distance between two point objects that can be separated on the moon with the eye whose pupil has a diameter of 5.5mm for λ=550 nm. We give distance earth-moon D=384 400km

$$\tan \theta \approx \theta = 1.22\frac{\lambda}{a} = \frac{1.22 \times 550 \cdot 10^{-9}}{5.5 \cdot 10^{-3}} = 1.22 \cdot 10^{-4}\ \text{rad}$$

The distance D between the two points is then:

$$D = \theta L = 1.22 \cdot 10^{-4} \times 3.844 \cdot 10^{8} = 46896.8\ \text{m}$$

Exercise II.3.6

A telescope has a mirror of diameter equal to 1m and is placed on a satellite at an altitude of 200 km to observe objects on earth at a wavelength of 400 nm.

What is the minimum distance between two objects on the earth's surface that it can distinguish?

Let's calculate the apparent angle θ under which the objects on earth are seen

$$\theta = \frac{1.22 \times 400 \, 10^{-9}}{1} = 4.9 \, 10^{-7} \text{ rd}$$

Minimum distance separating two objects located on the surface of the earth that it is able to distinguish:

$l = \theta \times L = 4.9 \, 10^{-7} \times 200 \, 10^{3} = 9.8 \, 10^{-2} \text{m} = 9.8 \text{ cm}$

Chapter III
Light interference

Two-wave light interference

III.1 TheoreticalRecall

III.1.1 History

Historically, light interference was discovered by the English scientist Thomas Young in 1801 used a single source which he split by an opaque paper in which he put two holes very close to each other. He observed that the two secondary sources gave on a screen interference. That is to say a phenomenon of bright bangs and dark bangs. This is how the phenomenon of light interference was demonstrated experimentally proving that light is a wave phenomenon.

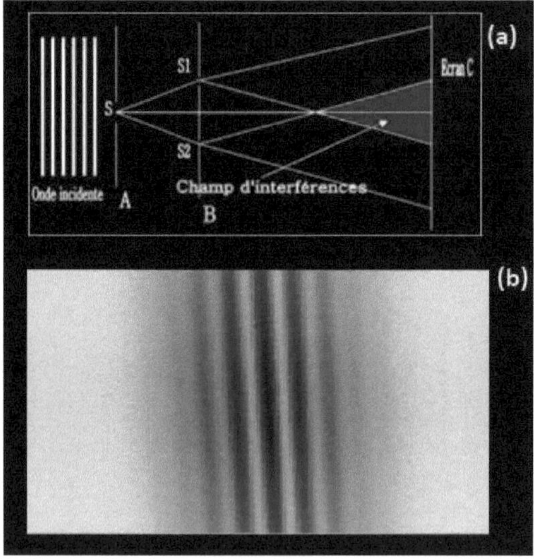

Fig.III.1 Phenomenon of light interference (a) T. Young's device. (b) Images of interference in visible white light on a screen

III.2 Theoretical reminder

III.2.1 Two-wave interference

We are interested in the field observed by a detector located at point M when two waves "1" and "2" arrive at M. Fig.1. These two waves are harmonic and have the same frequency.

Let P_1 and P_2 be two sources of same frequency emitting spherical waves

$$S_1 (M, t) = \frac{A_1 exp(-i(\omega t - kr_1 + \varphi_1))}{r_1} \qquad (III.1)$$

$$S_2 (M, t) = \frac{A_2 exp(-i(\omega t - kr_2 + \varphi_2))}{r_2} \qquad (III.2)$$

Because of the synchronization of these spherical waves, their superposition gives on the screen luminous interferences. These light interferences can be justified by a calculation as follows:

$$S(M, t) = S_1 (M. t) + S_2 (M, t) \qquad (III.3)$$

Thus we will have :

$$S(M, t) = \frac{A_1 exp(-i(\omega t - kr_1 + \varphi_1))}{r_1}$$
$$+ \frac{A_2 exp(-i(\omega t - kr_2 + \varphi_2))}{r_2} \qquad (III.4)$$

Figure III.1 gives the synoptic diagram of the two-wave interference experiment. We see that the point M on the screen is located at distances r_1 and r_2 from the two sources P_1 and P_2 :

$$r = r_{12} = D$$

Thus, the expression (III.4) becomes :

$$S(M, t) = \frac{A_1 exp(-i(\omega t - kr_1 + \varphi_1))}{D}$$
$$+ \frac{A_2 exp(-i(\omega t - kr_2 + \varphi_2))}{D} \qquad (III.5)$$

The light intensity is obtained by squaring the amplitude S(M, t):

$$I =_M \|(S(m,t))^2\| = \left|\frac{A_1}{D}\right|^2 + \left|\frac{A_2}{D}\right|^2 + \frac{A_1 A_2}{D^2} \lfloor exp(-i(kr_1 - kr_2 + \varphi_1 - \varphi_2)) - exp(-i(kr_2 - kr_1 + \varphi_2 - \varphi_1)) \rfloor$$
$$(III.6)$$

Using trigonometry we finally get :

$$I_M = \left|\frac{A_1}{D}\right|^2 + \left|\frac{A_2}{D}\right|^2 + \frac{A_1 A_2}{D^2} cos\langle k(r_2 - r_1) + \varphi_2 - \varphi_1 \rangle) \qquad (III.7)$$

This allows us to write this expression as a function of the intensities of the sources P_1 and P_2 as follows:

$$I_M = I_1 + I_2 + 2\sqrt{I_1 I_2}\, cos\langle k(r_2 - r_1) + \varphi_2 - \varphi_1 \rangle) \qquad (III.8)$$

III.2.2. Difference **in speed**

The step difference between the two optical paths is defined by :

$$\delta = S_1 M - S_2 M = r_2 - r_1 = \sqrt{D^2 + (x + \tfrac{a}{2})^2 + y^2} - \sqrt{D^2 + (x - \tfrac{a}{2})^2 + y^2}$$

$$= D\sqrt{1 + \tfrac{1}{D^2}((x + \tfrac{a}{2})^2 + y^2)} -$$

$$D\sqrt{1 + \tfrac{1}{D^2}((x - \tfrac{a}{2})^2 + y^2)} \qquad (III.9)$$

As $D \gg x$ et $D \gg y$, the limited expansion of expression (III.9) gives :

$$\delta = S_1 M - S_2 M = \frac{ax}{D} \qquad (III.10)$$

By replacing the relation (10) giving the difference of march in the expression of the intensity given by the expression (8) we will have :

$$I_M = I_1 + I_2 + 2\sqrt{I_1 I_2}\, cos\langle k(\tfrac{ax}{D}) + \varphi_2 - \varphi_1 \rangle) \qquad (III.11)$$

if we replace the wave vector $k=\dfrac{2\pi}{\lambda}$

$$I_M = I_1 + I_2 + 2\sqrt{I_1 I_2}\,cos\,\langle\frac{2\pi}{\lambda}\frac{ax}{D} +$$
$$\varphi_2 - \varphi_1) \qquad (III.12)$$

If we also assume that: $I = I_{12} = \dfrac{I_0}{2}$ and by posing $\varphi = \varphi_2 - \varphi_1 -$ we will have thus :

$$I = {}_M\frac{I_0}{2}\,(1 + 2cos(\frac{2\pi ax}{\lambda D} + \frac{\varphi}{2})\,)) \quad (III.13)$$
$$I_M = I_0\,cos^2\,(\frac{\pi ax}{\lambda D} + \frac{\varphi}{2}) \quad (III.14)$$

III.2.3 Interfrange

The intensity I_M varies sinusoidally and the intensity takes maximum and minimum values in a periodic way.

$I = I_{M0}$ we will have bright bangs if $cos^2\,(\frac{\pi ax}{\lambda D} + \frac{\varphi}{2})=1$ i.e. :

$$\frac{\pi}{\lambda}\frac{ax}{D} = m\pi \quad (III.15)$$

So:

$$x = m\,\frac{\lambda D}{a}$$
$$(III.16)$$

The interrange is given by the difference between two consecutive positions x_m and x_{m+1} thus :

$$i = x_{m+1} - x = m\frac{\lambda D}{a}$$
$$(III.17)$$

If we observe a number N of bangs of width L on a screen, then the inter-fringe is obtained by the following formula:

$$i = \frac{L}{N+1}$$
$$(III.18)$$

III.2.4 Two wave interference device: Fresnel mirrors

2.1. The Fresnel mirrors are a two-wave interference device. This device was designed by Fresnel to demonstrate the wave character of light. Two mirrors M_1 and M_2 are illuminated with a source S, making a very small angle α between them.
2.2. A point source S is placed at the distance $d = SO = 0.5$ m from the common edge of the mirrors. These give two images S_1 and S_2 of S located on the circle of center O and radius d.

2.3. The angle $S_1 OS_2$ is 2α and since α is very small, the distance
2.4. S S_{12} is :

$$S\ S_{12} = a = 2\alpha.d \ (III.19)$$

The beams coming from S undergo on M_1 and M_2 a reflection of the same nature. The two beams have a common part (in green). In this area, there is a superposition of two waves from the synchronous sources S_1 and S_2 : these waves interfere.

2.4. A screen is placed at E on the normal to S S_{12} passing through O. The interferences are studied by means of an optical device with a lens to magnify the obtained interferences.
2.5. We observe rectilinear bangs parallel to the edge of the mirrors. The central fringe is bright.   We pose :
$$D' = D + d.$$

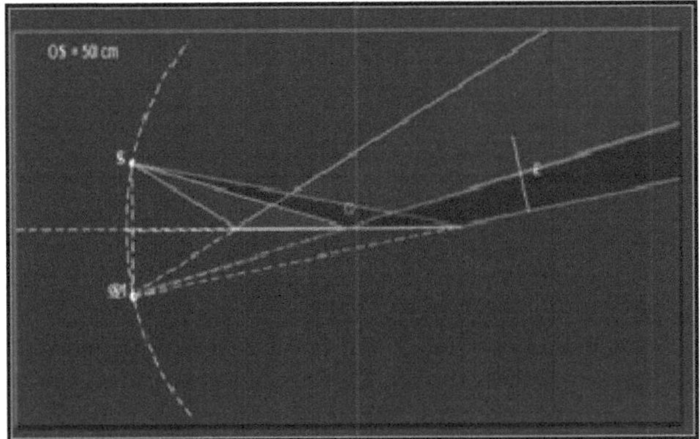

Fig.III.2 Two wave interference device Fresnel mirrors

The interfringe is therefore :

$$i = \frac{\lambda D\prime}{a} = \frac{\lambda(D+d)}{2\alpha d} = \frac{\lambda D}{2\alpha d} + \frac{\lambda}{2\alpha}$$
(III.20)

To improve the brightness, the point source can be replaced by a slit parallel to the edge of the mirrors.

3. Experimental device of the laboratory

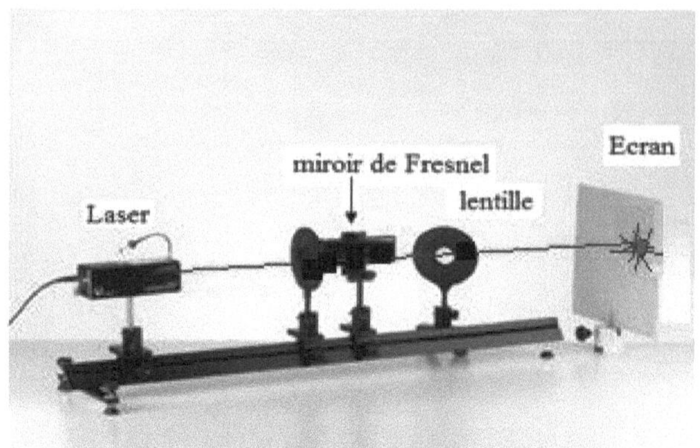

Fig.III.3 Interferential device of Fresnel mirrors

III.3 Applications to interference

III.3.1 Exercise 1:

We use the following setup in which Young's slits are separated by a=1.5mm and the interference bangs are observed on a screen located at a distance D=1.0 m. We can measure on the screen a set of 6 consecutive bangs on a width of 21 mm.

1. Explain the formation of interference on the screen. Do the interference bangs cover the entire screen? Explain qualitatively why one observes on the screen an alternation of luminous bangs and dark bangs, and not a continuum of light intensity.
2. Is the fringe in the center of the screen (y=0) bright or dark? Justify.
3. What is the interest in measuring the space corresponding to 6 bangs instead of one fringe?
4. . Recall the existing relationship between the interrange i and the quantities D, a and l the wavelength of the laser. Use this relation to determine the value of the wavelength.

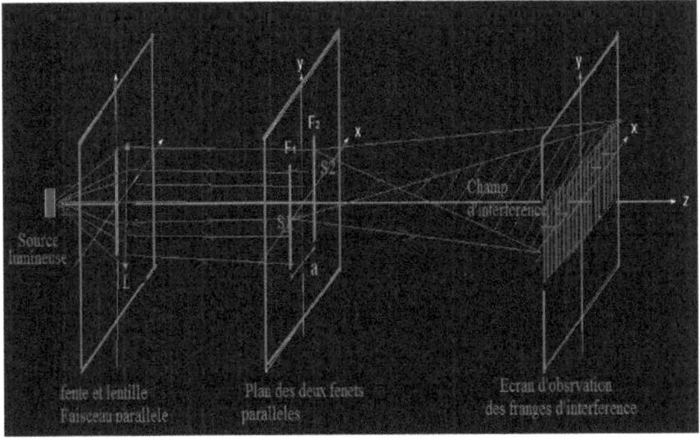

Fig. III.4 Experimental device for Young's slits

Solution exercise1

Answer to question III.3. 1

The sources are synchronous because they are duplicated from the same source S. Thus the two luminous vibrations coming from the sources S_1 and S_2 of the opaque screen having the same direction there is vectorial addition of the amplitudes of the two vibrations and thus interferences. Fig.III.4

Let φ be the phase shift in x fig.III.4 between the plane waves from the sources.

The two sources have the same amplitude A, the light intensity in x according to the geometry of the device used fig.III.4, this intensity is

$$I(x) = I_0 cos^2 \left(\frac{\pi a x}{\lambda D}\right) \text{ (III.21)}$$

$I(x) = I_{Max0}$ we will have bright bangs if $cos^2 \left(\frac{\pi a x}{\lambda D}\right) = 1$ i.e. :

$$\frac{\pi}{\lambda}\frac{a x}{D} = m\pi \text{ (III.22)}$$

$$\text{With } m = ... -2 \ -1 \ 0 \ +1 \ +2$$

The intensity I(x) takes maximum and minimum values according to the values of **m** whole and takes minimum values ie zero for the values of $x = m +$ **and this in a periodic way where the appearance of the fringes called bright fringes and where the intensity is zero I(x) = 0, we find the fringes called dark.**$\frac{1}{2}$ and this in a periodic way where the appearance of bright bangs called bright bangs and where the intensity is zero I (x) = 0, we find the bangs called dark.

The fact that the bangs appear on a part of the screen and do not cover it entirely is due to the field of interference limited by the region of meeting of the two light beams from S_1 and S_2 hatched region of fig. III.4

Answer to question III.3. 2

Since the fringe in the center of the screen corresponds to x=0, this will give the value **m=0** which is an integer value, which gives a bright fringe.

Answer to question III.3.3

Measuring several bangs instead of a single fringe comes back to the measurement system available in our experiment. By using a ruler whose accuracy is 1mm and which sometimes corresponds to the width of a fringe. So taking the length of several bangs at once is to reduce the error of measurement on the inter-fringe i. Similarly, on interferential systems as in a Michelson apparatus, we must count several bangs with a micrometer screw.

Answer to question III.3. 4

The inter-fringe **i** is given by the difference between two successive positions of two bright bangs

$$i= x_{m+1} - x_m = \frac{\lambda D}{a} \qquad\qquad (III.23)$$

So, given the numerical values of the problem; we can easily determine the wavelength λ

$$\lambda = \frac{ia}{D} \qquad\qquad (III.24)$$

First, the value of the interrange must be determined.

Either

$$i= \frac{L}{N+1}$$

$$i=\frac{21}{6+1}=3 \text{ mm}$$

So with a=1.5mm, D=1.0. m; we easily obtain

$$\lambda = \frac{ia}{D}=\frac{3*1.5}{1000}= 4.5 10^{-4} \text{ mm}=0.4500 \mu m$$

This wavelength corresponds to a blue color of the source used.

III.3.2 Exercise 2

Two identical slits of width a= 0.5mm are illuminated by a light source able to emit two radiations of respective wavelengths λ_1 and λ_2 . We observe on the screen located at a distance D = 2.00 m from the plane of the slits the system of bangs so that the distance separating the middle of the first bright fringe (the central fringe) and that of the tenth bright fringe is equal to 9.9 mm for the figure formed using the radiation of wavelength λ_1 . Switching radiation, we find that the radiation of wavelength λ_2 its ninth bright inter-fringe coincides with the eighth bright fringe due to the radiation of wavelength λ_1 .

1. What is the wavelength λ_1 of the radiation used?

1.a. Show a semifigure of the interference pattern and calculate the interrange i_1 for the radiation λ_1 .

1.b. Deduce the wavelength λ_1 .

2. a Calculate the distance separating the middles of the central fringe and the eighth bright fringe for radiation of wavelength λ_1 .

2.b. Calculate the interrange i_2 for radiation of wavelength λ_2 after schematizing the situation.

2.c Deduce the wavelength λ_2 .

Solution exercise III.4.2

The schematic of the Young device used in this experiment e rience is given in Fig.III.5. We first take the fringe system formed by the radiation of wavelength λ_1 .

1.a representation of the interference pattern obtained by the two Young's slits. Fig. III.5.

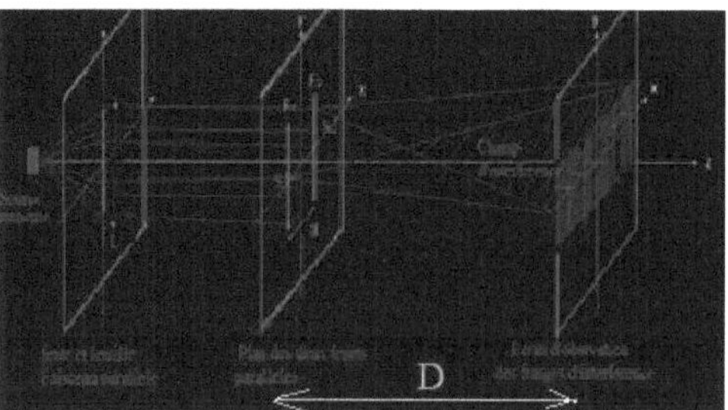

Answer to question 1.b

Calculation of the wavelength λ_1 .

The central fringe at z=0 is bright we will have the 9 bangs all bright so the inter-fringe is calculated by the relationship (II.18) is :

$$i =_1 \frac{L}{N+1} = \frac{9.9}{9+1} = 0.99 \text{ mm}$$

The wavelength is related to the interrange by the relation (II.24) or :

$$\lambda_1 = \frac{ia}{D} = \frac{0.99 \cdot 10^{-3} \times 1.2 \cdot 10^{-3}}{2} = 0.594 \cdot 10^{-6} \text{m} = 0.5940 \mu m$$

Answer to question 2.a

Calculate the distance separating the central and eighth bright fringe media for radiation of wavelength λ_1 .
The distance to the eighth fringe comprises N+1 intervals, i.e. 8 inter-fringes i_1 i.e.: for the representation see fig. III.6

$$d = (8+1) \, i = 8_1 \times i =_1 9 \times 0.99 = 8.91 \text{mm}$$

Answer to question 2b.
The inter-fringe i_2 is obtained as follows, the ninth fringe of λ_2 corresponds to the eighth fringe of λ_1 at the distance d=8.91 mm or :

$$(N+1) \, i =d_2$$
$$i =_2 \frac{d}{N+1} = \frac{8.91}{9+1} = 0.891 \text{ mm}$$

Answer to question 2c

Taking into account the relation (III.24) we deduce the wavelength λ_2 as follows:

$$\lambda_2 = \frac{i_2\, a}{D} = \frac{0.891\,10^{-3} \times 1.2\,10^{-3}}{2} = 0.5346\ 10^{-6}\,\text{m} = 0.5346\,\mu m$$

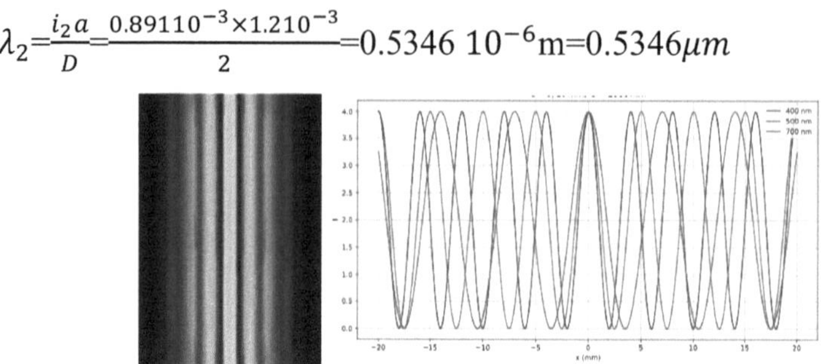

Fig.III.6: Image of a superposition of bangs of different colors and their graphic representations

II.3.4 Exercise 4

The Fresnel mirror opposite gives a source S two sources S_1 and S_2 corresponding to the images given by the two mirrors M_1 and M_2 making a small angle α. The source is placed at a distance d from the edge of the two mirrors. The screen is located at a distance D from this edge.

1. Express the expression of the interrange i as a function of the parameters: d, D, α and the wavelength λ.

We give: d=50cm and λ=632 nm.

2. Plot the graph i as a function of D for the following values of D :

D(cm)	50.	70	90	110	130	150	170	190	210
i (cm)	0.06	0.08	0.09	0.1	1.1	0.125	0.14	0.15	0.16

3. From the graph, deduce the angle α between the two mirrors M_1 and M_2

We give:d=50cand λ=632 nm

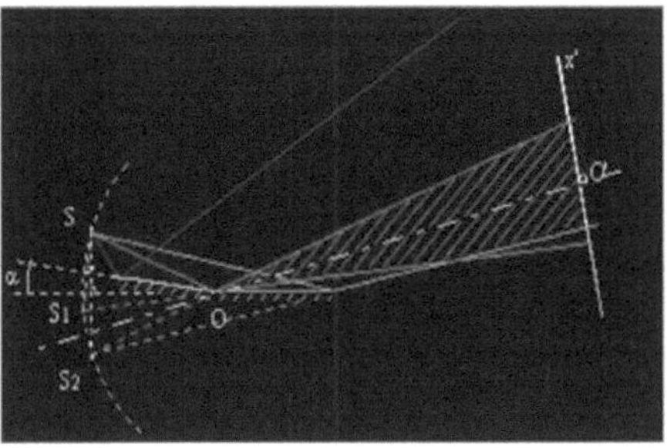

Fig.III.7 Device of the two Fresnel mirrors two-wave interference

We take the expression (III-17) which defines the interfrange as a function of the distance. Let :

$$i= x_{m+1} - x = m\frac{\lambda D}{a}$$

This expression has been demonstrated for interference bangs obtained by Young's slits. For Fresnel mirrors the distance sources - observation screen is D'=D+d, which gives for the inter-fringe i :

$$i = \frac{\lambda D\prime}{a} = \frac{\lambda(D+d)}{2\alpha d} = \frac{\lambda D}{2\alpha d} + \frac{\lambda}{2\alpha}$$

Plotting the graph

The plot of the graph is obtained point by point using the Origin software from which the line joining the maximum of points that intersects the axis of the inter-fringes and allows to deduce$\frac{\lambda}{2\alpha}$.=0.05 cm fig.III.8 Thus by using the data of the problem one calculates easily the angle$a$ is :

$$a = \frac{2\lambda}{0.0510^{-2}} = \frac{2\times63210^{-9}}{0.0510^{-2}} = 0.002528 \text{ rad} = 2.52810^{-3}\text{rad}$$

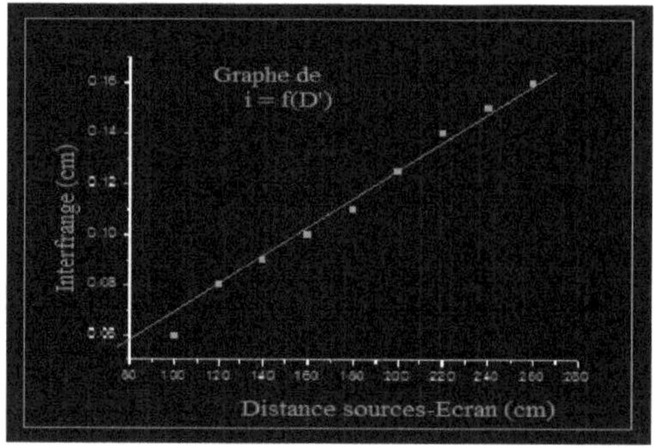

Fig.III.8 : Plot of the graph obtained experimentally thanks to the Fresnel mirrors experiment.

II.3.5 Exercise 5

The micrometric devices centered on E (Screen fig. III1(a), (b), (c)) and using light sources emitting two different colors makes it possible to give the observable interferences on the screen. We can therefore measure the interfringement by reading on the graph. as mentioned on fig. III .6

We give below several results corresponding to different positions D of the screen, that is to say we vary the distance D.and we take the distance 'a' between the two parallel slits is fixed, in all experiments, and equal to 0.56 mm

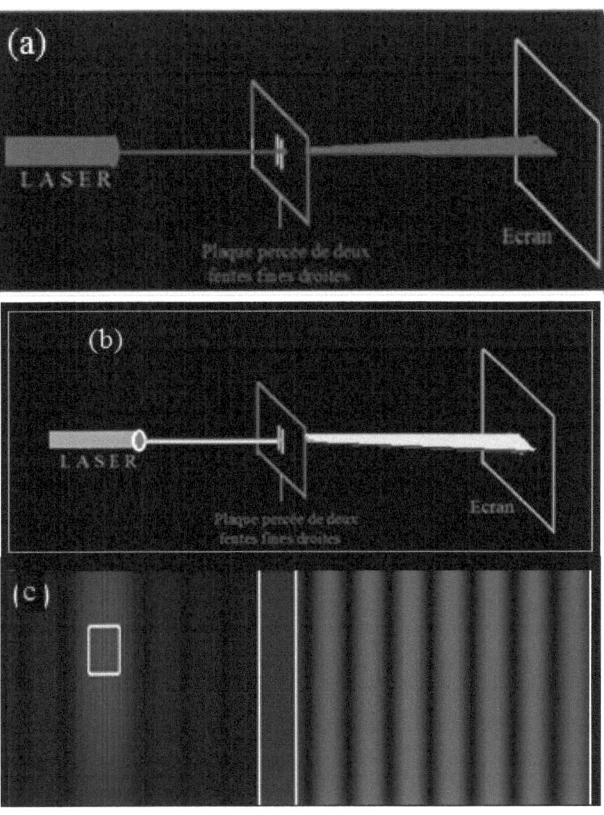

Fig.III.6 Schematics of two-wave interference devices

Result 2 : Wavelength λ_2 :(red color)

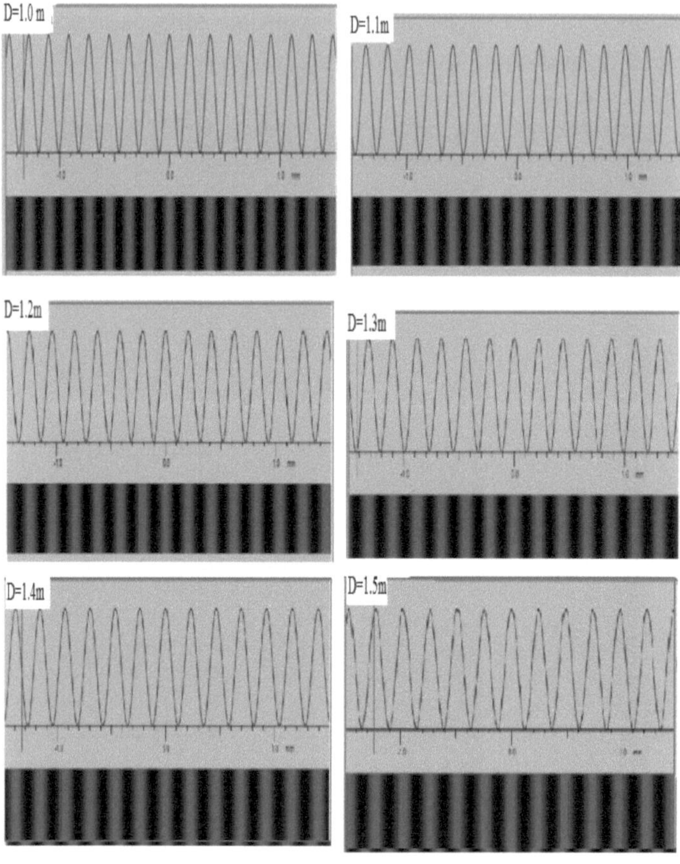

Fig.III.7 Two wave interference for red radiation

Result 3: : Wavelength λ₃ :(yellowish color)

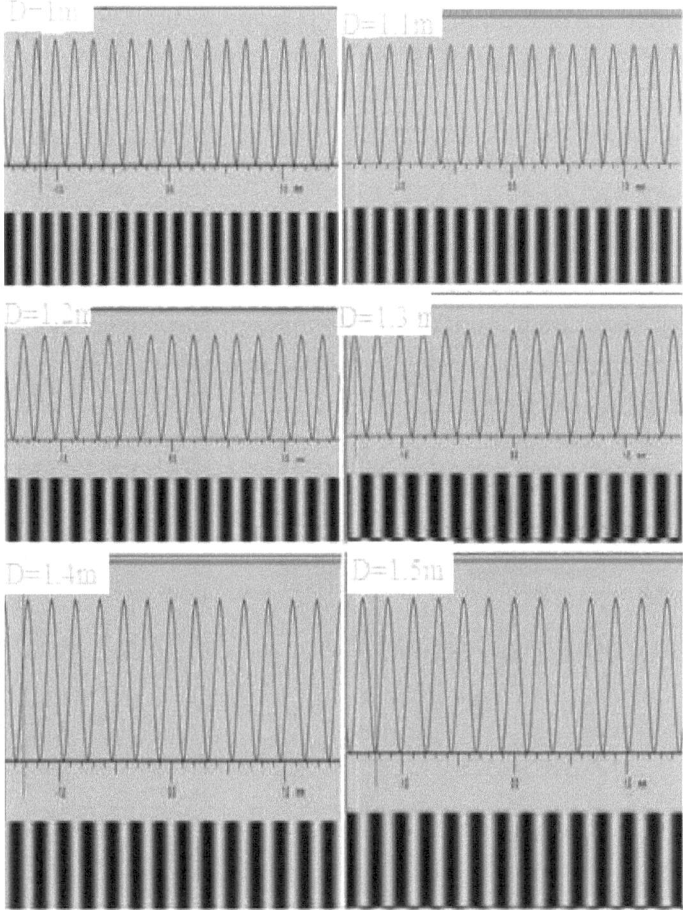

Fig.III.8 Two wave interference for   radiation

Questions
1. Determine for each color and for all distances the interrange i
2. Draw up a table of values showing the variations of the interfrange as a function of the distance D
3. Draw on a millimeter paper and, for the three colors, the variations of i as a function of D.
4. Deduce the values of the three wavelengths

5 Calculate the uncertainties on the calculated values
We give: $\Delta D = 1cm \ \ et \ \Delta i = 0.01mm$

Solution of the exercise II.3.5

1.Tableua of values
The table III.1 represents the values obtained of the different interferential sequences red yellow and green we summarize on this table the varaiation of the interfranges of warm color that is of each wavelength according to the distance to which is measured the interg=frange.

Table III.1 Interfringe values of the two wavelengths as a function of the distance D

D(m)	i_{Jaune} (mm)	i_{Rouge} (mm)
1.0	0.157	0.168
1.1	0.166	0.187
1.2	0.178	0.192
1.3	0.181	0.2
1.4	0.2	0.207
1.5	0.208	0.225

2. graphic representation

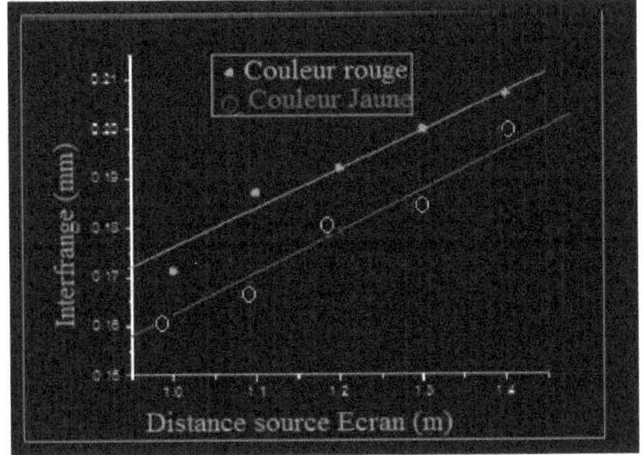

2. Determination of the two wavelengths

We can see that the graphical representation is a straight line passing through the origin, so we find a linear function in the form :

$$i= \frac{\lambda*D}{a}$$

And the ratio $\frac{\lambda}{a}$ is a constant equal to the slope of the line.

$$i= \frac{\lambda*D}{a} \quad , \quad \frac{\lambda}{a}= \frac{i}{D} \quad , \quad \lambda=\frac{i*a}{D}$$

$$a= 0.56mm$$

For the red color we will have to calculate the slope of the line:

$$\mathbf{p_R} =\frac{\lambda R}{a} = \frac{(0.225-0.168)10^{-3}}{(15-10)10^{-1}}=0.011410^{-2}$$

$$\lambda_R =0.5610^{-3} \times0.011410^{-2}=0.0638410^{-5} = 658.4nm$$

$$\mathbf{p_J} =\frac{\lambda j}{a} = \frac{(0.208-0.157)10^{-3}}{(15-10)10^{-1}}=0.010210^{-2}$$

$$\lambda_J =0..5610^{-3} \times0.0102010^{-2}=0.00571210^{-5}=571.2nm$$

λ_R and λ_J correspond well to the respective colors red and yellow on the visible spectrum of light.

Chapter IV
Light polarization

VI.1 Introduction to Light Polarization

This chapter deals with the *vector* character of electromagnetic waves. Transparent media will be considered isotropic. The word isoptropic means that light, which is an electromagnetic wave, has the same speed regardless of the direction taken by the light ray. However, there are media for which the optical properties depend on the direction it is the anisotropy.

The wave model is a vector model. For example, *Maxwell*'s equations in vacuum lead to "plane wave" solutions such as the trihedron $(\vec{E}$, $\vec{B}$, $\vec{k})$ is a direct trihedron. The light vector is none other than the *electric field* vector of the electromagnetic wave, it is the only one to which the usual receivers are sensitive (eye, photocell, etc.) and is therefore written :

$$\vec{E} = \vec{u} E_0 \cos(\omega t - \varphi)$$
(IV.1)

$\vec{u}$ is a unit vector designating a direction, E_0 is the amplitude of the electric field $\vec{E}$ and φ the phase at the origin.

IV.1.2 Unpolarized wave

A wave is said to be unpolarized or *natural* if in the plane of vibration, the field has no preferred direction. The components of the vibrating vector have no phase relationship . This change is related to the totally random character of the light emission. The configuration of the vibrating vector at time $t + \Delta t$ is completely different from that at time t, which is what characterizes natural or unpolarized light.fig.IV.1

IV.1.3 Polarized wave and polarization types

A wave is said to be polarized if the components of the electric field vector have a phase relationship. To describe the field, we place ourselves

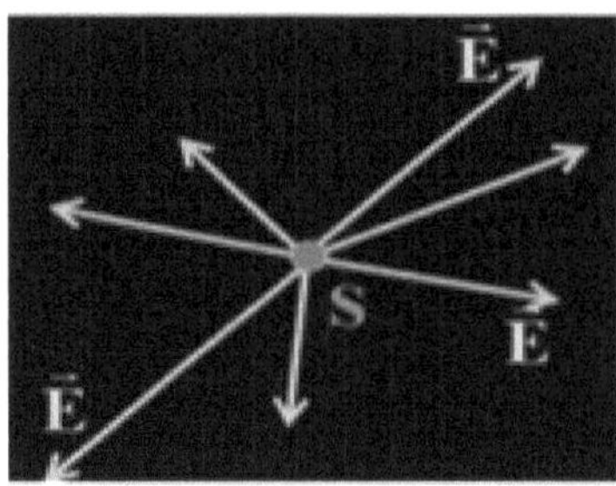

Fig. IV.1 *unpolarized or natural wave in the plane of vibration, the electric field has no preferred direction.*

in the xOy plane and we describe the evolution of the vector

$$\vec{E} \begin{cases} E_x = E_{0x}\cos(\omega t - \varphi_1) \\ E_y = E_{0y}\cos(\omega t - \varphi_2) \\ \quad E_z = 0 \end{cases}$$

(IV.2)

E_{0x} and E_{0y} are respectively the amplitudes of the components E_x et E_y. φ_1 and φ_2 are the phases at the origin of the two components E_x et E_y By proceeding to a change of variable taking into account the phase shift $\varphi = \varphi_2 - \varphi_1$ we will have the two components E_x et E_y which are written:

$$E_x = E_{0x}\cos\omega t$$
((IV.3)

$$E_y = \cos(\omega t - \varphi) = E_{0y}(\cos\omega t \cos\varphi + \sin\omega t \sin\varphi) \quad (IV.4)$$

We multiply (IV.3) by $\sin\varphi$ *et* we replace in (IV.4) $\cos\omega t$ by $\dfrac{E_x}{E_{0x}}$ we obtain the two equations :

$$\frac{E_x}{E_{0x}}\sin\varphi =\cos\omega t\ \sin\varphi$$
(IV.3')

$$\frac{E_y}{E_{0y}} -\frac{E_x}{E_{0x}}\cos\varphi= \sin\omega t\ \sin\varphi \qquad\qquad (IV.4')$$

Squaring (3') and (4') member to member and summing member to member we have:

$$\left(\frac{E_y}{E_{0y}}\right)^2+\left(\frac{E_x}{E_{0x}}\right)^2-2\frac{E_y}{E_{0y}}\frac{E_x}{E_{0x}}\cos\varphi=\sin\varphi^2 \qquad\qquad (IV.5)$$

This is the equation of an *ellipse*: the end of the field describes an ellipse. Depending on the value of φ, the ellipse is described in one trigonometric direction (right elliptical polarization) or in the other (left elliptical polarization).

Equation (IV.5) is the general equation allowing, according to the values of φ , to know all the existing types of polarization

Thus

1. Linear polarization for $\varphi=0$ or$\varphi=\pi$. Indeed the equation (IV 5) becomes :

$$E_y=\pm\frac{E_{0y}}{E_{0x}}E_x$$
(IV.6)

Equation (IV.6) gives a linear relationship between the E_y and the component E_x

2. Left and right circular polarization $\varphi=\pm\dfrac{\pi}{2}$

In these two cases equation (IV. 5) becomes :

$$\left(\frac{E_y}{E_{0y}}\right)^2 + \left(\frac{E_x}{E_{0x}}\right)^2 = 1$$

(IV.7)

This is the equation of a *circle of radius unity*: the end of the field describes a circle. Depending on the value of φ, the circle is described in a trigonometric direction (right circular polarization) or in the other (left circular polarization). The direction of rotation thus depends on the sign of sinφ.

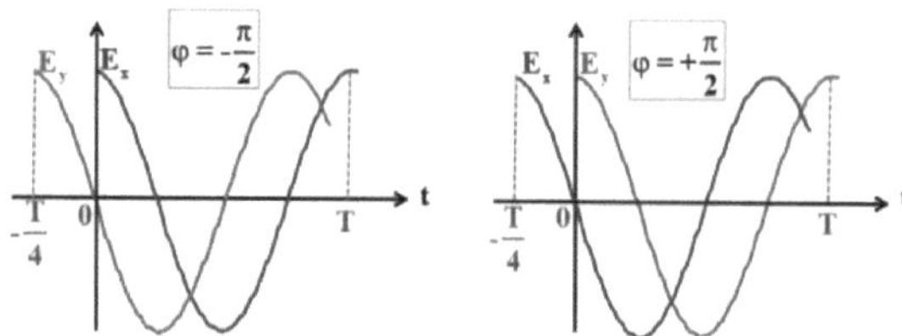

Fig. IV.2 Circular polarization case: For $\varphi=-\dfrac{\pi}{2}$ the component E_x component is in advance of t=$\dfrac{T}{4}$ compared to the component E_y where T is the rotation period. If $\varphi=+\dfrac{\pi}{2}$ the component E_x is in delay of t=$\dfrac{T}{4}$.

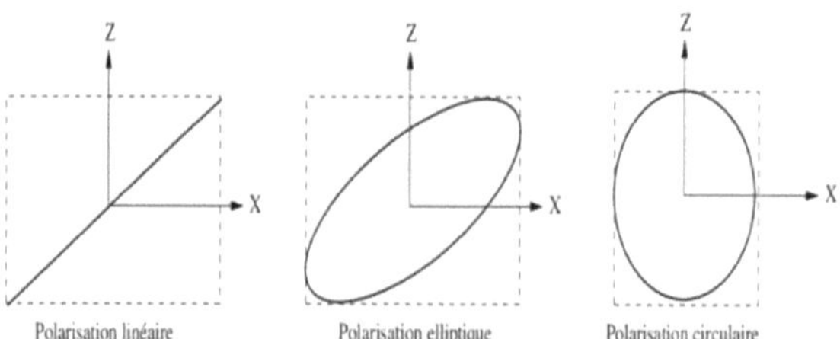

Fig. IV.3: Diagram of the types of polarization

IV.1.4. Jones' formalism

IV.1.4.1 Electric field and Jones vector

The expressions of the components E_x et E_y of the electric field vector given in expression (IV.2) can be written in exponential notation as follows:

$$\vec{E}\begin{cases}E_x = E_{0x}e^{-i(\omega t-\varphi_x)}\\E_y = E_{0y}e^{-i(\omega t-\varphi_y)}\end{cases}=e^{i\omega t}\begin{cases}E_x = E_{0x}e^{i\varphi_x}\\E_y = E_{0y}e^{i\varphi_y}\end{cases} \qquad (IV.8)$$

$$=e^{i\omega t}\begin{cases}E_x = E_{0x}\\E_y = E_{0y}e^{i(\varphi_y-\varphi_x)}\end{cases}=\begin{cases}E_p\\E_s\end{cases}=[J]$$

(IV.9)

$e^{i\omega t}$ is called the propagator term and $[J]$ is the Jones vector.

Ep : the electric field is parallel to the plane of incidence, in the plane of the optical table

Es : The electric field is perpendicular (from the German word *senkrecht*), to the plane of the table.

IV.1.4.2 Types of polarization according to Jones

- Linear (or rectilinear): if $-=k\varphi_y\varphi_x\pi$ with $k\in N$, if $E_x =0$ or $E_y =0$. For linear polarization, three cases are possible
It is noted L_H (horizontal linear) for the field is horizontal parallel to the plane of the table

$$L_H=\begin{Bmatrix}0\\E_0e^{i\omega t}\end{Bmatrix}=\begin{bmatrix}0\\1\end{bmatrix} \qquad (VI.10)$$

If it is noted L_V : (the field is vertical to the plane of the table L_V

$$=\begin{Bmatrix}E_x\\E_y\end{Bmatrix}=\begin{Bmatrix}0\\E_0e^{i\omega t}\end{Bmatrix} \qquad (VI.11)$$

If we note it L_θ : The field is directed according to an angle θ with respect to Ox that is

$$L_\theta=\begin{pmatrix}cos\theta\\sin\theta\end{pmatrix} \qquad (VI.12)$$

A right-handed circular polarized wave is noted R_H from the two English words (Right Handed). It is written as follows:

$$R_H = \frac{1}{\sqrt{2}} \begin{pmatrix} 1 \\ -i \end{pmatrix} \tag{VI.13}$$

For a left-handed circular wave we note L_H from the two English words (Left Handed)

$$L_H = \frac{1}{\sqrt{2}} \begin{pmatrix} 1 \\ i \end{pmatrix} \tag{VI.14}$$

The delay blades, better known as wave blades or phase shifters, while transmitting the light beam, plays the role of a modifier of the state of polarization of light, transmit without attenuating or changing the direction of the beam. All they do is perform a phase delay of the component E_y in relation to the component E_x of the electric field vector $\vec{E}$ vibrating electric field vector of the light wave.

These phase shifter blades are made from a birefringent material, i.e. having two refractive indices n_x and n_y along two orthogonal directions, generally E_x and E_y . They are made of a crystalline material, generally quartz. The polarized light undergoes a modification according to these two orientations represented in general by the directions of the two components E_x and E_y of the electric field vector$\vec{E}$. They separate the incident unpolarized light into its parallel and orthogonal components. The polarized light is dephased according to the retarder plate which will be either a wave plate or a half wave plate or finally a quarter wave plate. Fig. IV.4

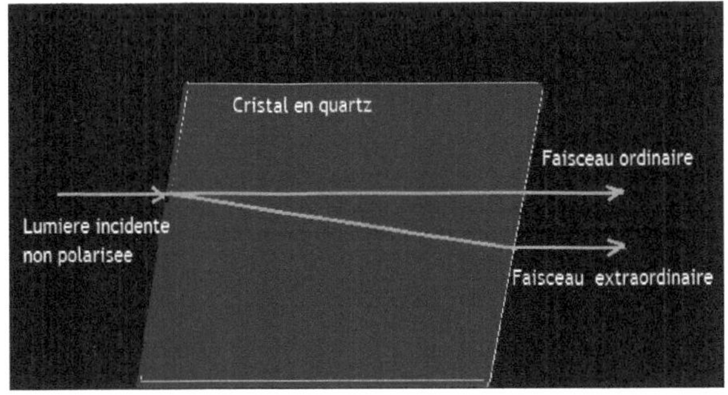

F.IV.4 Representative schematic of a phase shifting plate

The most commonly used retarder or phase shifter blades in experiments are a quarter wave $\lambda/4$ blade, a half wave $\lambda/2$ blade and a λ wave blade. Fig. IV.5 Other values of the delay will be used in some applications. Such as the internal reflection of a prism that causes a phase change between components, which can be a problem. A compensating delay blade can restore the desired polarization.

IV .1.3 Half-wave plate (*HWP*)

- thickness e=λ /2, delay φ =π
 - has two *neutral lines* at 90° from each other.

- transforms an incident rectilinear polarization into its symmetric with respect to the neutral line. In other words: a rectilinear polarization rotated byφ with respect to the neutral lines rotates by 2φ . Example: forφ =45° the emerging polarization is perpendicular to the incident polarization.

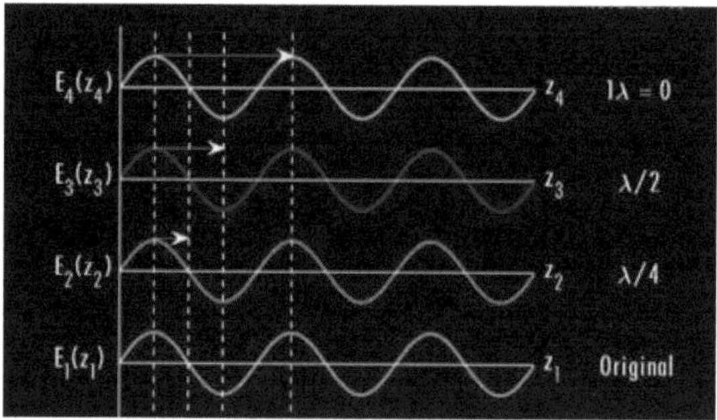

Fig. IV.5 Representative schematic of the type of delay and phase shifters of an original wave.

It is known that the wave blades induce phase shifts between the E_y component and the E_x component of the polarized electric field. These phase shifts are calculated as follows:

$\varphi = -\varphi_y \; \varphi_x = 2\pi\delta/\lambda$ (VI.15)
With
$$\delta = (n_y - n_x)e \qquad\qquad (VI.16)$$
δ; difference de marche n_y et n_x; indices de refraction respectifs suivant Oy and Ox and 'e' the thickness of the blade.

Thus, the wave blades are noted as follows: "For a blade that gives a difference of march :
- λ called wave blade, the phase shift is $0°$ it is noted e :

1. Wave blade

$$L_0 = \begin{bmatrix} 1 & 0 \\ 0 & -1 \end{bmatrix} \qquad (IV.17)$$

3. Half-wave blade

$* \dfrac{\lambda}{2}$ called half-wave blade, the phase shift being 45° it is noted :

$$L_{45°} = \begin{bmatrix} 0 & 1 \\ 1 & 0 \end{bmatrix} \qquad\text{(IV. 18)}$$

4. Quarter wave blade

$* \dfrac{\lambda}{4}$ called quarter-wave plate, the phase shift being 90° it is noted :

$$L_{90°} = \begin{bmatrix} -1 & 0 \\ 0 & 1 \end{bmatrix} \qquad\text{(IV. 19)}$$

Other phase shifts are also known, their expressions are the following

$$L_{22.5°} = \frac{1}{\sqrt{2}} \begin{bmatrix} 1 & 1 \\ 1 & -1 \end{bmatrix}, \qquad\text{(IV. 20)}$$

$$L_{67.5°} = \frac{1}{\sqrt{2}} \begin{bmatrix} -1 & 1 \\ 1 & -1 \end{bmatrix}, \qquad\text{(IV.21)}$$

IV.1.3. Law of Malus
IV.1.3.1 Principle of the Malus law

Etienne Louis Malus (1775-1812) was the first physicist and mathematician to perform an experiment on the polarization of natural light. His experiment is described as follows:

A converging lens is illuminated in natural light, i.e. unpolarized, using a source placed at its object focus. The parallel beam emerging from the lens illuminates a polarizer P. The light is then linearly polarized (rectilinear polarization). Just behind P we place another polarizer called analyzer (it analyzes the light polarized by

P). We note θ the angle between the axes of P and A. fig.IV.5 Thus if E_1 is the electric field that has passed through P, the field that passes through A is the projection of E_1 on the axis of A.
We obtain

$$E=E_1\cos\theta \qquad\qquad (IV.22)$$

The measurable light intensity I would be the square of the amplitude of E or :

$$I=E_1^2\cos^2\theta \qquad\qquad (IV.23)$$

If we put $I =_0E_1^2$ then we have :

$$I=I_0\cos^2\theta \qquad\qquad (IV.24)$$

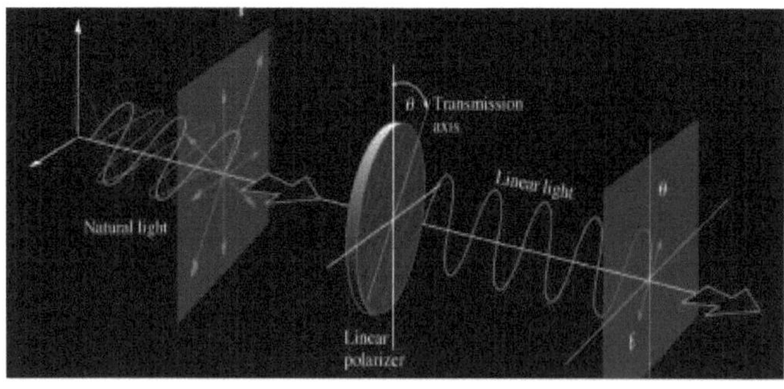

Fig. IV.5 Schematic diagram of the principle of polarized wave propagation

IV.1. 3.2 Experiment and plot of the Malus curve

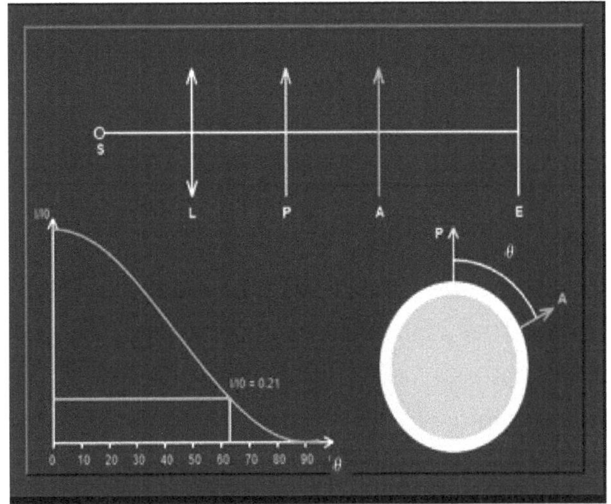

Fig. IV.6: Diagram of the Malus experiment and plot of the Malus curve as a function of the angle θ that the analyzer makes with the axis of the polarizer.

IV.2 Application exercises on polarization

Exercise 1

Describe the polarization state of the following waves and represent the electric field vector $\vec{E}$ for t=0 ;T/4 ;T/2,and 3T/4 with T , the ^period of the light vibrations.

1°) $\vec{E} = \vec{e_x} E_0 \sin \omega t + \vec{e_y} E_0 \sin (\omega t + \pi/2)$

2°) $\vec{E} = \vec{e_x} E_0 \sin \omega t + \vec{e_y} E_0 \sin (\omega t + \pi/4)$

3°) $\vec{E} = \vec{e_x} E_0 \sin \omega t + \vec{e_y} E_0 \sin (\omega t + \pi/2)$

4°) $\vec{E} = \vec{e_x} E_0 \sin \omega t - \vec{e_y} E_0 \sin (\omega t + \pi/2)$

Exercise 2
Two linearly polarized waves are in phase but have different amplitudes at x=0 :

$$\vec{E_1}=\vec{e_x}A_1 \cos \omega t + \vec{e_y} B_1 \cos\omega t$$
$$\vec{E_2}= \vec{e_x} A_2 \cos \omega t + \vec{e_y}B_2 \cos\omega t$$

Show that $\vec{E}=\vec{E_1} +\vec{E_2}$ is also linearly polarized.

Exercise 3

Two waves are circularly polarized right and left at x=0:
$$\vec{E_1}= \vec{e_x} E_0 \cos \omega t + \vec{e_y} E_0 \sin \omega t$$
$$\vec{E_2}= \vec{e_x} E_0 \cos (\omega t + \varphi) - \vec{e_y}E_0 \sin(\omega t+\varphi)$$

Show that their sum $\vec{E}=\vec{E_1} +\vec{E_2}$ form a linear polarization and find the direction of this polarization.

Exercise 4

An elliptically polarized wave is written at x=0 :

$$\vec{E}= \vec{e_x} 4\cos \omega t +\vec{e_y}\sin \omega t$$

Show that this wave can be decomposed into two waves, one circularly polarized and the other linearly polarized.

IV.4 Exercises on the Jones formalism

Exercise 5

a) Describe the type of polarization of the electric field vector in the following Jones representation and represent the vector $\vec{E}$ in an orthonormal plane Oxy

$$\begin{bmatrix}1\\0\end{bmatrix} \; ; \; \begin{bmatrix}0\\1\end{bmatrix} \quad ; \; \frac{1}{\sqrt{2}}\begin{bmatrix}1\\1\end{bmatrix} \; ; \; \begin{bmatrix}1\\i\end{bmatrix} ; \begin{bmatrix}1\\-i\end{bmatrix}$$

b) Describe the polarization state of the following two Jones vectors:

$$\begin{bmatrix} 1+i \\ 1-i \end{bmatrix} and \begin{bmatrix} 1-i \\ 1+i \end{bmatrix}$$

c) The waveplate introduces a phase shift ϕ, give the value of ϕ in each case;

$$\begin{bmatrix} 1 & 0 \\ 0 & -i \end{bmatrix}\begin{bmatrix} i & 0 \\ 0 & 1 \end{bmatrix}$$

Exercise 6

We wish to describe a propagation of a plane light wave, monochromatic, in matrix form.

A polarizer acts on the electric field of a wave by an operator noted P. Similarly, we note L the matrix corresponding to the action operator of a birefringent plate.

1. Write the Jones vector for a wave with rectilinear and circular polarization.

2. Write the matrix $P(\theta)$ for a polarizer whose axis makes an angle θ with the direction Ox in the wave plane $(O; x, y)$.

3. Write the matrix $P\varphi$ corresponding to the operator of the action of a birefringent blade introducing a phase delay φ between the E_x et E_y components of the electric field. Explain.

Exercise 7

A beam of white light passes through a set of two crossed polarizers P and A. Between P and A, a birefringent blade is placed, with the faces perpendicular to the incident beam and the optical axis Ox arranged relative to P as shown in Figure IV.7. The thickness of the blade is e = 0.26 mm; the ordinary index n_o and extraordinary index

n_e are such that $n_e - n_o = 0.00165$. Consider the wavelength range between 420 and 580 nm

1. What are the wavelengths for which no light emerges from A ?

2. What are the wavelengths for which the light coming out of A has the same intensity as that coming out of P ?

3. What can we conclude about the use of such a set-up?

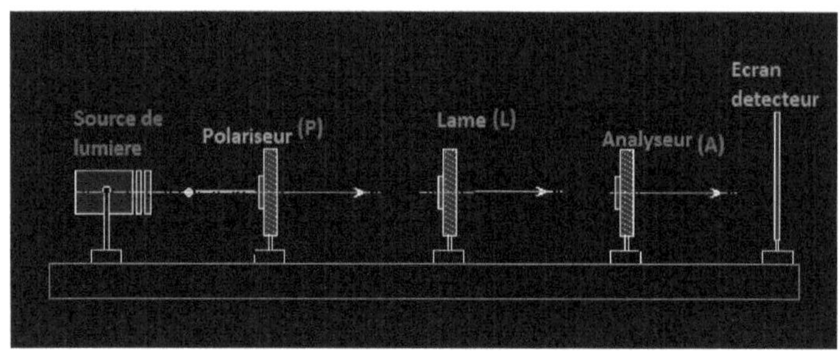

Fig. IV 7: Schematic of a system of polarizer, wave blade and analyzer

5. Solutions to the exercises

<u>Solution exercise 1</u>

1°) $\vec{E} = \vec{e_x} E_0 \sin \omega t + \vec{e_y} E_0 \sin (\omega t + \pi/2)$ implies $E_x = \sin \omega t$ and $E_y = \sin (\omega t + \pi/2) = \cos \omega t$ thus: $E_x^2 + E_y^2 = 1$ which gives a circular polarization.

2°) $\vec{E} = \vec{e_x} E_0 \sin \omega t + \vec{e_y} E_0 \sin (\omega t + \pi/4) = \vec{e_x} E_x + \vec{e_y} E_y$ we have $\sin (\omega t + \pi/2) = \sin \omega t \cos \pi/2 + \cos \omega t \sin \pi/2 = \cos \omega t = E_y / E_0$ (2)

$\sin \omega t = E_x / E_0 \ .....$ (3)

Replacing (2) ,(3) and in 1 and squaring we will have :

$$E_x^2 / E_0^2 + E_y^2 / E_0^2 = 1$$

The same reasoning allows us to deduce:

2°) We have an elliptical polarization with an angle $\pi/4$ with the Ox axis.

3°) Linear polarization $E_y / E_x = 1$

4°) Linear polarization $E_y / E_x = -1$

Solution Exercise 2

$\vec{E}=\vec{E_1}+\vec{E_2}== (\vec{e_x} A_1 \cos \omega t + \vec{e_y} B_1 \cos\omega t) +(\vec{e_x}A_2 \cos \omega t + \vec{e_y} B_2 \cos\omega t)=\vec{e_x}A_1 + A_2) \cos \omega t \; \vec{e_y}(B +B_{12}) \cos \omega t$. This gives us

$E_y = ((A_1 + A_2)/ (B +B_{12})) E_x$

We will therefore have a linear polarization.

Solution Exercise 3

By adding memebre to memmbre and taking the ratio of the components of the resulting field we find :

$E_x /E_y = (\cos \omega t +\cos (\omega t + \varphi)) / (\sin\omega t- \sin(\omega t+\varphi))$ and taking into account the trigonometric relations below We have: $E_x /E_y = \tan \varphi$ which expresses a linear polarization

Solution Exercise 4

We write:

$\vec{E}= \vec{e_x}A\cos \omega t+\vec{e_y} B\sin \omega t=\vec{e_x}A\cos \omega t++ \vec{e_y} B\sin \omega t++ \vec{e_y}A\sin \omega t-+ \vec{e_y}A\sin\omega t$
$=(\vec{e_x}A\cos \omega t+\vec{e_y} A\sin\omega t) +\vec{e_y}(B\sin \omega t -A\sin\omega t)$

The first expression in blue represents a circular polarization while the one in green is a linear polarization

Solution Exercise 5

This is the Jones representation for the vector $\vec{E}$ in an orthonormal plane Oxy.

Let us consider the first case J= $\begin{bmatrix}1\\0\end{bmatrix}$ in this case, we will have :

$$\begin{Bmatrix}E_x\\E_y\end{Bmatrix}=\begin{Bmatrix}E_0e^{i\omega t}\\0\end{Bmatrix}=\begin{Bmatrix}1\\0\end{Bmatrix} \text{, which gives}: E_x = E_0e^{i\omega t} \text{ et } E_y = 0$$

The wave vibrates in the direction of the Ox axis and is therefore polarized along the unit vector $\vec{e_x}$

Let us consider the first case J= $\begin{bmatrix}0\\1\end{bmatrix}$ in this case, we will have :

$$\begin{Bmatrix}E_x\\E_y\end{Bmatrix}=\begin{Bmatrix}0\\E_0e^{i\omega t}\end{Bmatrix}=\begin{bmatrix}0\\1\end{bmatrix} \text{which gives}: E_y = E_0e^{i\omega t} \text{ et } E_x = 0$$

The wave vibrates in the direction of the Oy axis and is therefore polarized along the unit vector $\vec{e_y}$

Let us consider the first case J=$\frac{1}{\sqrt{2}}\begin{bmatrix}1\\1\end{bmatrix}$ in this case, we will have :

$$\begin{Bmatrix}E_x\\E_y\end{Bmatrix}=\frac{1}{\sqrt{2}}\begin{bmatrix}1\\1\end{bmatrix}=L_\theta=\begin{pmatrix}cos\theta\\sin\theta\end{pmatrix}=\begin{pmatrix}cos45°\\sin45°\end{pmatrix}$$

The wave vibrates in the direction making an angle of 45 with respect to Ox it is therefore polarized along the unit vector $\vec{u}$ such that :

$$\vec{u}=\cos 45°\vec{e_x} + \sin 45°\vec{e_y}$$

Let us consider the first case J=$\begin{bmatrix}1\\i\end{bmatrix}$ in this case, we will have :

$$\begin{Bmatrix}E_x\\E_y\end{Bmatrix}=\begin{Bmatrix}E_0e^{i(\omega t)}\\E_0e^{i(\omega t++\varphi)}\end{Bmatrix}= e^{i(\omega t)}\begin{Bmatrix}1\\e^{i\varphi}\end{Bmatrix} =\begin{bmatrix}1\\i\end{bmatrix}$$

We will have $e^{i\varphi}=i$, i.e.

$$\cos\varphi + i\sin\varphi = i$$

We will have:

$$\varphi = \frac{\pi}{2}$$

The polarization is then left circular (see course)

Let us consider the first case $J=\begin{bmatrix} 1 \\ -i \end{bmatrix}$ in this case, we will have :

$$\begin{Bmatrix} E_x \\ E_y \end{Bmatrix} = \begin{Bmatrix} E_0 e^{i(\omega t)} \\ E_0 e^{i(\omega t++\varphi)} \end{Bmatrix} = e^{i(\omega t)} \begin{bmatrix} 1 \\ e^{i\varphi} \end{bmatrix} = \begin{bmatrix} 1 \\ -i \end{bmatrix}$$

We will have $e^{i\varphi}=i$, that is

$$\cos\varphi + i\sin\varphi = -i$$

We will have:

$$\varphi = -\frac{\pi}{2}$$

The polarization is then right circular (see course)

Answer to question 5 b

Let be the following Jones vector: $\begin{bmatrix} 1 + i \\ 1 - i \end{bmatrix}$

It can be written in the form

$$\begin{bmatrix} 1 + i \\ 1 - i \end{bmatrix} = \begin{bmatrix} e^{i\frac{\pi}{4}} \\ e^{-i\frac{\pi}{4}} \end{bmatrix} = e^{i\frac{\pi}{4}} \begin{bmatrix} 1 \\ e^{-i\frac{\pi}{2}} \end{bmatrix} = e^{i\frac{\pi}{4}} \begin{bmatrix} 1 \\ -i \end{bmatrix} = \begin{bmatrix} 1 \\ -i \end{bmatrix}$$

It is therefore a right circular polarization

Let be the following Jones vector: $\begin{bmatrix} 1 + i \\ 1 - i \end{bmatrix}$

$$\begin{bmatrix} 1 + i \\ 1 - i \end{bmatrix} = \begin{bmatrix} e^{-i\frac{\pi}{4}} \\ e^{i\frac{\pi}{4}} \end{bmatrix} = e^{-i\frac{\pi}{4}} \begin{bmatrix} 1 \\ e^{i\frac{\pi}{2}} \end{bmatrix} = e^{i\frac{\pi}{4}} \begin{bmatrix} 1 \\ i \end{bmatrix} = \begin{bmatrix} 1 \\ i \end{bmatrix}$$

It is therefore a left circular polarization

A wave plate (WP) - a neutral line rotated by an angleθ with respect to the x axis - a plate of thickness e introduces a delay

$$\varphi = -\varphi_y \quad \varphi_x = 2\,e/\pi\lambda$$

Such as:

$$L_\varphi = \begin{bmatrix} 1 & 0 \\ 0 & e^{-i\varphi} \end{bmatrix} = \begin{bmatrix} 1 & 0 \\ 0 & -i \end{bmatrix}$$

By comparing the two members, we can write
$e^{-i\varphi} = \cos\varphi - i\sin\varphi = -i$
Then we will have $\varphi = \dfrac{\pi}{2}$

The difference in speed is then $\varphi = \dfrac{\pi}{2} = \dfrac{2\pi\delta}{\lambda}$

Thus:

$$\delta = \frac{\lambda}{4}$$

It is a quarter-wave blade

In the second case of exercise 5c we will have :

$$L_\varphi = \begin{bmatrix} 1 & 0 \\ 0 & e^{-i\varphi} \end{bmatrix} = \begin{bmatrix} 1 & 0 \\ 0 & i \end{bmatrix}$$

By comparing the two members, we can write
$e^{-i\varphi} = \cos\varphi - i\sin\varphi = i$
Then we will have $\varphi = -\dfrac{\pi}{2}$

The difference in speed is then $\varphi = -\dfrac{\pi}{2} = \dfrac{2\pi\delta}{\lambda}$

Thus:

$$\delta = -\frac{\lambda}{4}$$

This is another quarter-wave blade in the opposite direction of the first.

Solution exercise 6

Response to question 6.1

$$\vec{E}\begin{cases} E_x = E_{0x}e^{-i(\omega t-\varphi_x)} \\ E_y = E_{0y}e^{-i(\omega t-\varphi_y)} \end{cases} =e^{i\omega t}\begin{cases} E_x = E_{0x}e^{i\varphi_x} \\ E_y = E_{0y}e^{i\varphi_y} \end{cases}$$

$$=e^{i\omega t}\begin{cases} E_x = E_{0x} \\ E_y = E_{0y}e^{i(\varphi_y-\varphi_x)} \end{cases} = \begin{cases} E_x \\ E_y \end{cases} = [J]$$

In the case where the wave is rectilinearly polarized. We must have a linear relationship between the two components of the electric field is :

$$[J] = \begin{cases} E_x \\ E_y \end{cases} = \begin{cases} cos\theta \\ sin\theta \end{cases} , \theta \text{ is the angle that the electric field vector}$$

makes with the polarizer axis.

Response to question 6.2

The matrix of a polarizer whose axis makes an angle θ is written as follows:

$$P_\theta = \begin{pmatrix} cos\theta \\ sin\theta \end{pmatrix}$$

Response to question 6.3

To know the expression J' of the Jones matrix of an element J whose axes make an angle φ with the incident electric field, we apply the rotation matrices

$$J'=R(-\varphi)JR(\varphi)$$

With

$$R(\varphi)=\begin{vmatrix} cos\varphi & sin\varphi \\ -sin\varphi & cos\varphi \end{vmatrix}$$

If the polarization is linear along Oy ,we write the Jones vector as follows : $J=\begin{bmatrix} 0 \\ 1 \end{bmatrix}$ and if we apply the matrix of the dephasing blade we will have .

$$J' = \begin{vmatrix} cos\varphi & -sin\varphi \\ sin\varphi & cos\varphi \end{vmatrix} \begin{bmatrix} 0 \\ 1 \end{bmatrix} \begin{vmatrix} cos\varphi & sin\varphi \\ -sin\varphi & cos\varphi \end{vmatrix}$$

For $\varphi = \pi$

$$J' = \begin{vmatrix} -1 & 0 \\ 0 & -1 \end{vmatrix} \begin{bmatrix} 0 \\ 1 \end{bmatrix} \begin{vmatrix} -1 & 0 \\ 0 & -1 \end{vmatrix} = \begin{bmatrix} 0 \\ -1 \end{bmatrix}$$ effectively the polarization direction is reversed

For $\varphi = \dfrac{\pi}{2}$

$$J' = \begin{vmatrix} 0 & -1 \\ 1 & 0 \end{vmatrix} \begin{bmatrix} 0 \\ 1 \end{bmatrix} \begin{vmatrix} 0 & 1 \\ -1 & 0 \end{vmatrix} = \begin{bmatrix} -1 \\ 0 \end{bmatrix}$$ there is a rotation of $\varphi = \dfrac{\pi}{2}$ the polarization of the electric field is along Ox.

$$\vec{E} = \begin{cases} E_x = E_{0x} cos\varphi \cos \omega t \\ E_y = E_{0y} sins\varphi \cos \omega t \end{cases}$$

We thus obtain an elliptically polarized wave of axes the neutral lines of the blade.

If $\varphi = \pm \dfrac{\pi}{4}$ we obtain a circular polarization, right $\varphi = +\dfrac{\pi}{4}$ and left $\varphi = -\dfrac{\pi}{4}$. Thus to produce a circularly polarized light, it is necessary to introduce on the optical path a quarter-wave plate with un déphasage $\varphi = \pm \dfrac{\pi}{4}$

Solution to exercise 7

Response to question 7.1

The polarizer P and the analyzer A are crossed by assumption, which means that no wavelength can pass through A. If we interpose a birefringent plate, it automatically introduces a phase delay θ; such that :

$$\theta = \frac{2\pi\delta}{\lambda} = \frac{2\pi(n_e - n_\circ)e}{\lambda}$$

Thus Malus' law for the intensity of light emerging from A is written as follows:

$$I = I_0 cos^2\theta$$

the wavelengths for which no light emerges from A are such that I=0 i.e.

$$cos^2\theta = 0$$

Thus

$$\theta = \frac{2\pi(n_e-n_\circ)e}{\lambda} = (2m+1)\frac{\pi}{2}$$

What allows to draw the expression of the wavelengths by which no light emerges from A; Let/

$$\lambda = \frac{2\pi(n_e-n_\circ)e}{=(2m+1)\frac{\pi}{2}} = 4\frac{(n_e-n_\circ)e}{(2m+1)}$$

For m=0

$$\lambda_0 = 4\frac{0.00165\times0.2610^3}{(1)} = 171.610^{-9}$$

$\lambda_0 = 171.610^{-9}$ is not part of the given spectrum

For m=1

$$\lambda_1 = 4\frac{0.00165\times0.2610^3}{(2+1)} = 57210^{-9} = 572 \text{ nm}$$

$\lambda_1 = 572$ nm is included in the range of the given spectrum

For m=2

$$\lambda_2 = 4\frac{0.00165 \times 0.2610^3}{(4+1)} = 343.210^{-9}$$

$\lambda_2 = 343.2$ nm is also not included in the defined light spectrum. For m> 2 all wavelengths are outside the spectrum of the light used in this experiment.

In this case the light intensity emerging from the analyzer A is equal to that entering through the polarizer P.

We will have

$$I_0 = I_0 \, cos^2\theta$$

Or :

$$cos^2\theta = 1$$

Or/

$$\cos\theta = \pm 1$$
$$\theta = \frac{2\pi(n_e - n_\circ)e}{\lambda} = m\pi$$

This gives

$$\lambda = \frac{2\pi(n_e - n_\circ)e}{m\pi} = \frac{2(n_e - n_\circ)e}{m}$$

For m=1 we will have :
λ_1 =858nm which is not included in the spectrum used in this experiment.

For m=2 we have .
$$\lambda_2 = 429\text{nm}$$
This wavelength λ_2 =429nm is included in the range of the light spectrum used in this experiment

This setup can be used to select a single color for a pass or a block, it can be used as an optical fil

Chapter V
Propagation of light in transparent media

V.1 Introduction

V.2 Transparent media

A transparent medium is a place in which light propagates. The light is thus subject to propagate in the vacuum. Other transparent media are also suitable places in which light propagates easily like air, glass, water, plexiglass and precious stones. We can meet isotropic media where the light propagates the electric field polarizes is delayed but other transparent media are said to be anisotropic because of the presence of directions pressing different refractive indices and gives the electric field a phase delay. In both cases if the medium is homogeneous, the light propagates in a straight line

V.3 Opaque media

Despite the opacity of a medium, science has been able to demonstrate that there is a class of light wavelengths capable of passing through the opaque material without undergoing deformation, and allowing the transmission of information. For the demonstration, the researchers projected the image of a constellation through a white powder made of zinc oxide nanoparticles. They were able to recover the image on a sensor without major deformation despite a reduced light intensity

V.2 Fresnel coefficients

V.2.1 Introduction

Augustin Jean Fresnel (1788-1827), demonstrated certain relationships giving the coefficients of reflection and transmission of refractions of media through two media to describe the phenomena of reflection and refraction of electromagnetic waves at the interface between two media, whose refractive indices are respectively n_1 and n_2 . These coefficients, called Fresnel coefficients, express the relationship between the amplitudes of reflected and transmitted waves relative to the amplitude of the incident wave.

Thus it denoted the coefficient of reflection in amplitude by *r* and the coefficient of transmission in amplitude by *t* deduced from the respective ratios between the reflected field E_r and the transmitted field E_t compared to the incident electric field E_i . Thus we write :

$$r = \frac{E_r}{E_i} \tag{V.1}$$

and

$$t = \frac{E_t}{E_i} \tag{V.2}$$

Figure IV.1 represents the direction of propagation of the electric field E of the electromagnetic wave propagating between the two media of index n_1 and n_2.

In general, these coefficients depend on :

1. the dielectric constants of the input and output media, ε_1 and ε_2

2. the frequency f of the incident wave

3. angles of incidence $i_i = i_1$ and of refraction-transmission $i_t = i_2$,

4. the polarization of the waves. This leads to an eventual polarization of an initially unpolarized wave.

They are obtained by considering the continuity relations at the interface of the tangential components of the electric and magnetic fields associated with the wave.

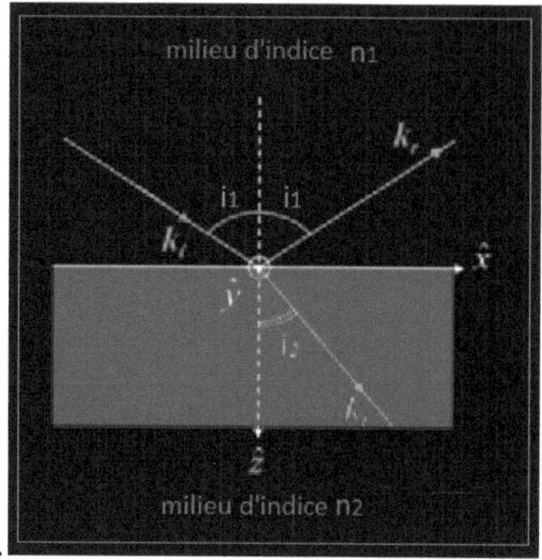

Fig. V.1 : Diagram of the reflection-transmission of a plane wave during an index jump

Let's consider two media, with different refractive indices, separated by a plane interface. The incident light wave is a plane wave, with wave vector $\vec{k}$, and of pulsation ω.

The Fresnel coefficients calculated here are valid only under the following assumptions about the media:

- They are linear, homogeneous and isotropic.

The Fresnel coefficients depend on the polarization of the electromagnetic field, we consider in general 2 cases:

V.2.2 Case of electrical transverse waves

Transverse electric (TE): the incident electric field is polarized perpendicular to the plane of incidence, the magnetic field is contained in the plane of incidence. Fig. V.2

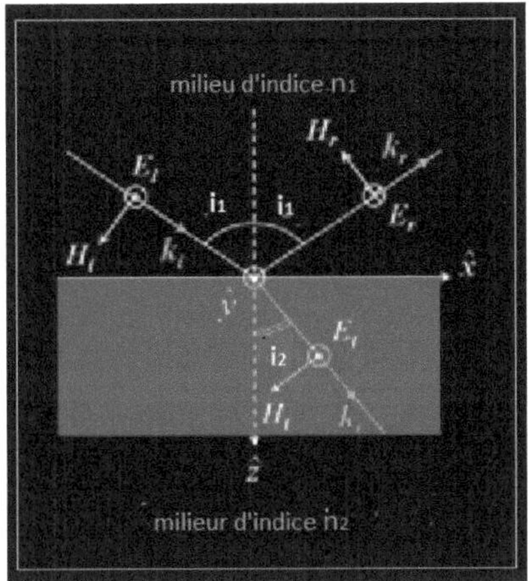

Fig. V.2 : Diagram of the reflection-transmission of a plane wave with $\vec{E}$ perpendicular to the plane of incidence

Let us consider an electromagnetic plane wave:

$$\vec{E} = E \, \exp j(\vec{k}\vec{r} --\omega t)\vec{e_y} \qquad (\text{V.3})$$

where E represents complex amplitude

Thus we deduce from the Maxwell-Faraday equation seen in the manuscript electromagnetism I:

$$\vec{H} = \frac{1}{\omega\mu}(\vec{k}\wedge\vec{E}) \quad (\text{V.4})$$

If we assume that the incident wave is polarized perpendicular to the plane of incidence, we have :

$$\overrightarrow{E_x}=$$

$$E\,\overrightarrow{e_x} \qquad\qquad\qquad\qquad (V.5)$$

At the interface between the two media, and from Maxwell's laws, we have continuity relations between the different components of the electric and magnetic fields. Thus, the tangential components of the electric field are maintained. The electric fields are already parallel to the interface, so we have

$$E_i = E_r + E_t$$
(V.6)

As far as the magnetic field is concerned, it is the normal components that are retained. As $\vec{E}$ is perpendicular to $\vec{B}$ the incident, transmitted and reflected magnetic fields are in the plane of incidence and we have, by projection on the axis (Oz) :

$$(B_i - B_r).\cos(i_1)=B_t\cos(i_2) \qquad\qquad\qquad (V.7)$$

where i_1 et i_2 are respectively the angle of incidence and refraction.

Moreover, we have seen in the electromagnetic manuscript I that the Maxwell-Faraday equation allows us to establish that :

$$B{=}n\frac{E}{c} \qquad\qquad\qquad\qquad (V.8)$$

We can then write:

$$n_1(E_i - E_r).\cos(i_1)=n_2 E_t\cos i_2 \qquad\qquad (V.9)$$

Dividing equation (V.9) by E_i we will have

$$n_1(1 - r_\perp).\cos(i_1)=n_2 t_\perp\cos i_2$$
(V.10)

With $r_\perp = \dfrac{E_r}{E_i}$ and $t_\perp = \dfrac{E_t}{E_i}$

By introducing, for each medium, the dispersion relation , $k = n\dfrac{\omega}{c}$ obtains the Fresnel coefficients according to the characteristics of the incidence (n_1 ,θ_1) and refraction (n_2 ,i_2) :

By replacing E_t by its expression in (1), we can write :

$$n_1 (E_i - E_r). \cos i_1 = n_2 (E_i - E_t) \cos i_2 \qquad (V.11)$$

Dividing equation (V.11) by E_i

$$n_1 (1 - r_\perp). \cos(i_1) = n_2 (1 - t_\perp) \cos i_2$$
(V.12)

and finally gives the reflection and transmission coefficients, called Fresnel coefficients:

$$r_\perp = \frac{E_r}{E_i} = \frac{n_1 \cos i_1 - n_2 \cos i_2}{n_1 \cos i_2 + n_2 \cos i_2} \qquad (V.13)$$

$$t_\perp = \frac{E_t}{E_i} = \frac{2 n_1 \cos i_1}{n_1 \cos i_1 + n_2 \cos i_2}$$
(V.14)

V.2.3 Case of magnetic transverse waves

Transverse magnetic waves (TM): we will have the incident magnetic field is polarized perpendicular to the plane of incidence, the electric field is contained in the plane of incidence. Fig.IV.3

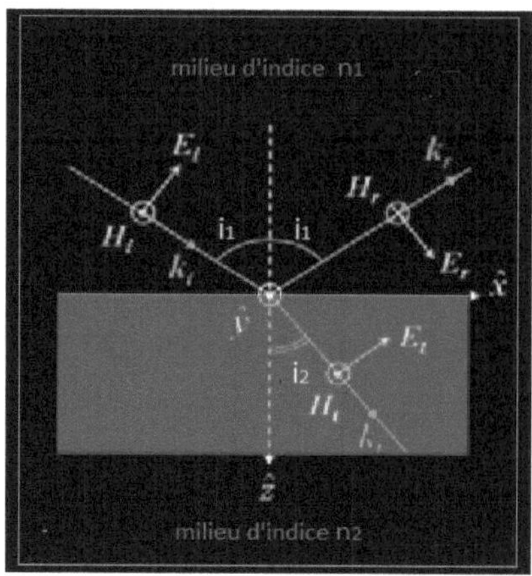

Fig. V.3: Schematic diagram of the reflection-transmission of a plane wave with the magnetic field perpendicular to the plane of incidence

By introducing the dispersion relation $k = n\dfrac{\omega}{c}$, for each medium, the Fresnel coefficients as a function of incidence (n_1 ,θ_1) and refraction (n_2 ,θ_2) are written t as follows "

$$H_i - H_r = H_t \tag{V.15}$$

This gives using the dispersion relation:

$$k_1(1-r_{/\!/}) \cos i_1 = k_2 t_{/\!/}) \tag{V.16}$$

Or by using the dispersion relation:

$$n_1(1-r_{/\!/})\cos i_1 = n_2 t_{/\!/} \tag{V.17}$$

With the two coefficients corresponding to the electric field $\vec{E}$ parallel to the plane of incidence that we will note :

$$r_{/\!/} = \frac{E_r}{E_i} \quad \text{and} \quad t_{/\!/} = \frac{E_t}{E_i} \tag{V.18}$$

On the other hand, we will have for the electric field contained in the plane of incidence the following relationship:

$$E_i \cos i_1 + E_r \cos i_1 = E_t \cos i_2 \tag{V.19}$$
$$(1+r_{/\!/})\cos i_1 = t_{/\!/}\cos i_2 \tag{V.20}$$

By combining the two relations (V.17) and (V.20) we can easily deduce the two Fresnel coefficients for the transverse magnetic field as follows:

$$r_{/\!/}=\frac{E_r}{E_i}=\frac{n_2 \cos i_1 - n_1 \cos i_2}{n_2 \cos i_1 + n_1 \cos i_2} \qquad (V.21)$$

and

$$t_{/\!/}=\frac{E_t}{E_i}=\frac{2n_1 \cos i_1}{n_2 \cos i_1 + n_1 \cos i_2} \qquad (V.22)$$

V.3 Application to Fresnel coefficients

Exercise 1

Represent the Fresnel coefficients and Fresnel factors for the following successive refractive indices

1. $N_1 = 1$ and $N_2 = 1.39$
2. $N_1 = 1$ and $N_2 = 1.50$
3. $N_1 = 1$ and $N_2 = 1.60$
4. Note for each case the Brewster angle

Solution questions for exercise 1

The Fresnel formulas translate the boundary conditions of the electric and magnetic fields when a light wave passes through a dioptre separating two media of index n_1 and n_2 .

We must distinguish the case of the electric field vibrating in the plane of incidence (magnetic transverse waves with r⊥ and t⊥) from the case of the electric field vibrating normally to the plane of incidence (electric transverse waves with r// and t//). The transmitted and reflected energies are proportional to the squares of the amplitudes (R = r^2) (T = t^2).. The program used to represent all these parameters designed by Mans university, allows to draw the curves **r=f(i), t=g(i) and**

R=F(i)

as well as **T=G(i)**

A vibration perpendicular to the plane of incidence is always 180° out of phase with the incident vibration. The reflected vibration changes direction with respect to the incident beam. The abrupt 180° change in phase occurs for the Brewster incidence (when the reflected beam is extinguished: we then have :

$$i_1 + i_2 = \pi / 2.$$

Note that the most important parameter is R(i) because it corresponds to the energy reflected by the dioptre. In optical systems, we seek to minimize (anti-reflection layers) or increase (dielectric mirrors)

Figure V .4 represents the coefficients), $r_\perp, t_\perp, r_{/\!/}$ $t_{/\!/}$ and the Fresnel factors $R\perp = r_\perp^2$, $R\| = r_{/\!/}^2$ and $T\perp = t_\perp^2$, $T\| = t_{/\!/}^2$ for the values of the respective refractive indices from Exercise 1.

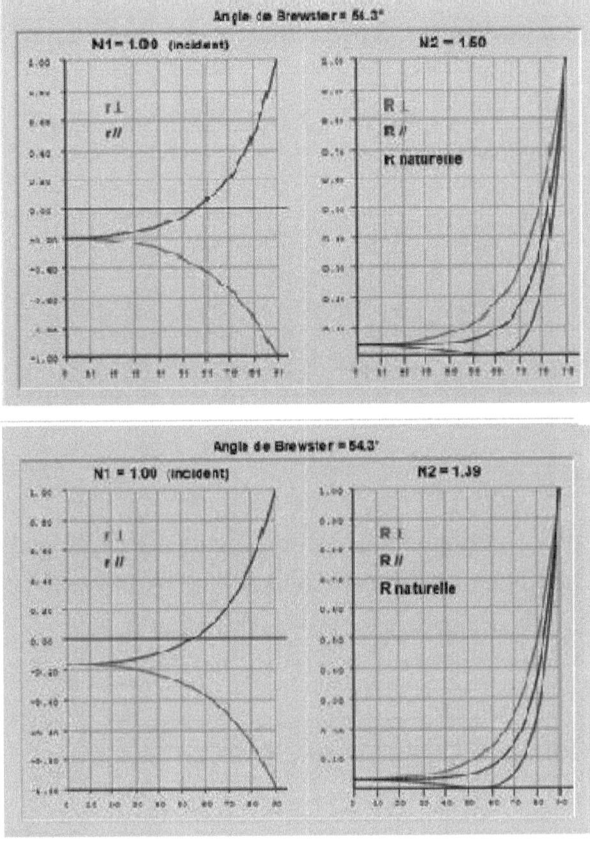

Fig.. V.4 Plot of Fresnel coefficients and factors

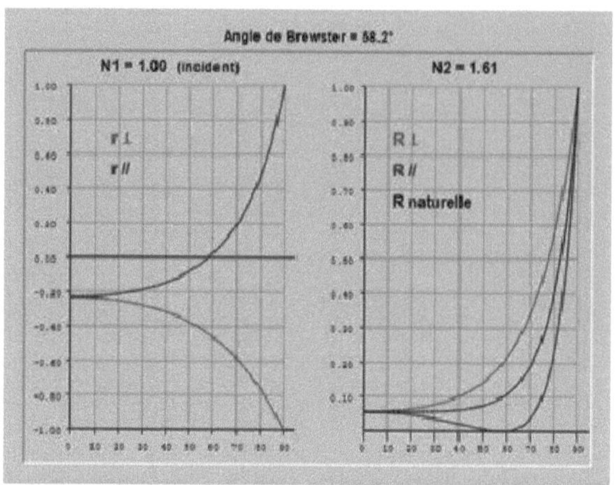

Exercise 2

The reflection and transmission coefficients are expressed only as a function of i_1 and i_2 of the indices n_1 and n_2 . Analyze the behavior of $r_\perp$ and $t_\perp$ for an incidence close to the normal.

Case n $_1$> n$_2$

Recall the formulas giving $r_\perp$ and $t_\perp$ are :

$$r_\perp = \frac{n_1 \cos i_1 - n_2 \cos i_2}{n_1 \cos i_2 + n_2 \cos i_2}$$

$$t_\perp = \frac{2n_1 \cos i_1}{n_1 \cos i_1 + n_2 \cos i_2}$$

For an incidence close to the normal these two expressions become :

$$r_\perp = \frac{n_1 - n_2}{n_1 + n_2}$$

$$t_\perp = \frac{2n_1}{n_1 + n_2}$$

Case n $_1$< n$_2$

This is the case of a reflection to a more refractive medium such as air / glass. In this case the relations giving the Fresnel coefficients are written as follows:

$$r_\perp = \frac{sin(i_2-i_1)}{sin\ (i_2+i_1)}$$

$$r_{/\!/} = \frac{tang(i_2-i_1)}{tang\ (i_2+i_1)}$$

We also deduce the reflection factors as follows

$$R\perp = r_\perp{}^2 = \left[\frac{sin(i_2-i_1)}{sin\ (i_2+i_1)}\right]^2$$

$$,\ R\| = r_{/\!/}{}^2 = \left[\frac{tang(i_2-i_1)}{tang\ (i_2+i_1)}\right]^2$$

Figure V.5 shows the graphical representation of the Fresnel coefficients and factors.

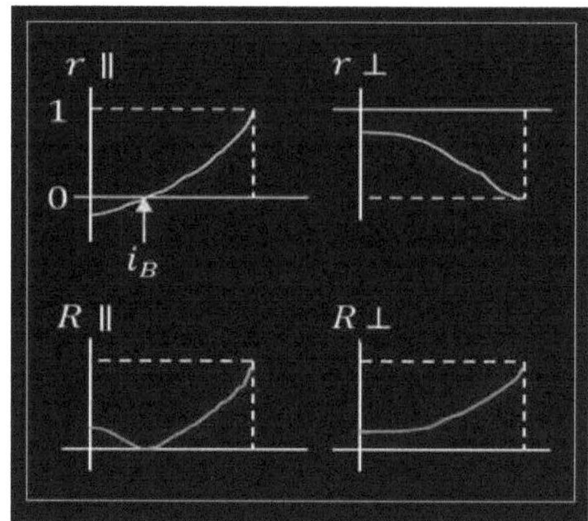

Fig. V.5: Graphical representation of Fresnel coefficients and reflection factors

Exercise 3

The reflected light is partially polarized while its polarization will be total when (r∥=R∥=0) i.e. at Brewsterian incidence.
1. Defining the Brewster's impact
2. Calculate in this case the reflected energy
3. Is the light in this case partially or totally polarized

Response to question 3.1
1. Brewer's incidence

We defined the Brewsterian incidence i_B for which r∥=0.
 This meant that after reflection, the parallel component of a natural light disappeared and that only its perpendicular component was reflected. The Brewsterian reflection allows to obtain a light polarized perpendicular to the plane of incidence. The Snell-Descartes formula allows us to easily predict the Brewster angle if we know the refractive indices n_1 and n_2 of the two media. Let the Snell-Descartes law applied figure V.6.

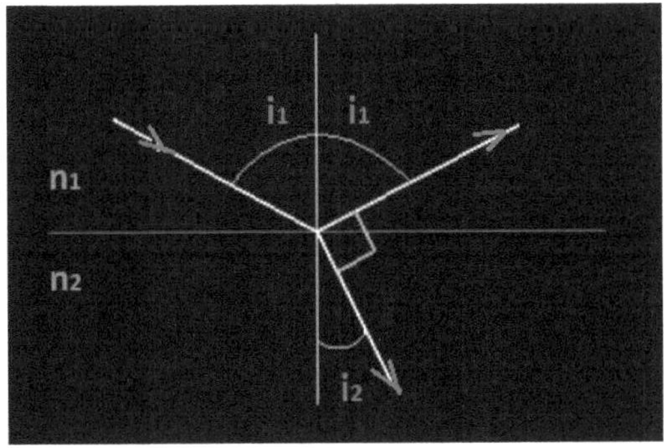

Fig. V.6 Brewster angle formed by the angle of incidence i_1

We will have:

$n_1 \sin i_1 = n_2 \sin I_2$

$\sin I_2 = \sin(\frac{\pi}{2} - i_1) = \cos i_1$

We will have :

$\text{Tang } i_1 = \frac{n_2}{n_1}$

Then let :

$i_1 = i_B = \text{actang } \frac{n_2}{n_1}$

2 Response to question 3.2

Let's calculate the reflection coefficients for the wave that reflects in the Brewsterian incidence case. We know that r∥ = 0, so R∥=0

Thus we will have :

$$r_\perp = \frac{\sin(i_1 - i_2)}{n\sin(i_1 + i_2)} = \frac{\sin i_1 \cos i_2 - \sin i_2 \cos i_1}{n_1 \cos \theta_2 + n_2 \cos \theta_2}$$

By dividing the two terms of the expression by :

$$cos i_2 \cos i_1$$

We will have:

$$r_\perp = \frac{\tan g\ i_1 - \tan g i_2}{\tan g i_1 + \tan g i_2}$$

On the other hand, we have

$$i_1 + i_2 = \frac{\pi}{2}$$

This allows for:

$$\tan g\ i_1 = \frac{1}{\tan g\ i_2}$$

With :

$$\tan g\ i_1 = \frac{n_2}{n_1} = n$$

Thus we obtain :

$$r_\perp = \frac{\frac{1}{n} - n}{\frac{1}{n} + n} = \frac{1-n^2}{1+n^2}$$

with

$$R\perp = r_\perp{}^2 = \left(\frac{1-n^2}{1+n^2}\right)^2$$

Response to question 3.3

The natural light is formed by two perpendicular vibrations of equal amplitudes and coherent. They can be considered perpendicular and parallel to the plane of incidence. From the point of view of reflection of the intensity, we must write :

$$R = R/\!/ + R\perp.$$

At the Brewster angle $R\| = 0$ therefore, all the light energy is included in $R\perp$. It is therefore completely polarized.

General conclusion

The light can be considered as the physical agent indispensable to the vision. In this manuscript we have defined the light as an electromagnetic wave, that is to say a propagation of electric and magnetic fields, vector quantities. In a small area around a point far from the source, we have considered the amplitude of the light wave constant and the direction of the electric field constant; only the phase varies significantly. Natural light (sunlight, lamps, candles) is unpolarized. Light is then treated as a scalar wave passing through isotropic media. For the human eye sensitive to the range of wavelengths 400nm-750nm, ie in the spectral region where our courses and our exercises are particularly focused, and that we call the field of visible. In this area, the frequencies are of the order of some 10^{14} Hz, the periods are of the order of some 10^{-15} s. The eye has its maximum visual acuity in the yellow green 560nm.

The study that has been made has therefore included a description of light as a wave. These properties are described by observable phenomena. We have seen that light is an electromagnetic wave with a dual aspect: It is an electromagnetic field that varies in space and time, so light is part of electromagnetic waves. This is its wave aspect. It is also a flow of particles or photons which gives light the corpuscular aspect.

Unlike the geometrical optics treated in optics I, optics II describes interference and diffraction phenomena that occur when the sources are coherent with each other. The phenomenon of light polarization is the process of converting unpolarized light into polarized light. Light in which the particles vibrate in a preferred direction or geometric shape is called polarized light. The Fresnel coefficients, express the relationship between the amplitudes of the reflected and transmitted waves in relation to the amplitude of the incident wave. They evaluate the optical behavior of optical multilayers, from the simplest of them, the blade with parallel face, to the most complex: anti-reflection layers, interference filters.

Finally, where electrons and copper links are limited to a few gigabits per second, photons are the only ones able to transport information with exemplary quantity, efficiency and speed.

This manuscript will certainly be followed by a third fascicle in which will be treated the interferential systems which translate interference phenomena resulting from the superposition, which can be constructive or destructive, of waves presenting different phase differences. We can mention the interferential systems by amplitude division which are of great practical importance such as: the Michelson interferometer, the Mach-Zehnder interferometer, the Sagnac interferometer and the Fabry-Pérot interferometer.

We will also give an idea on their uses in fiber optics ensuring the functions of transport and spatial filtering of the optical signal in an interferometer and integrated optics which ensures complementary functions such as spatial filtering; and optimization of interferometric coding.

Based on this type of analysis and thanks to the mastery of the basic elements of optics in general, science has been able to develop Integrated Optics components of increasing complexity that have been validated in the laboratory and then in everyday life on the internet and on the ethernet.

References

[1] Hadj Ali Bakir, optical communication, Master course handout, Hassiba Ben Bouali University Chlef (2019)

[2] Alicia Maréchal, course on polarization, University of Nantes Studocu (2021)

[3] Mohamed Lotfi, Wave Optics - Detailed course and summary of courses and problems posed to the competitions, École Normale Supérieure de Marrakech (2020)

[4] H. Djelti , Optoelectronics, mimeo of Master course, Abou Bakr Belkaid University of Tlemcen (2017)

 [5] A.Bouzid, wave optics, course TD and TP Moulay Ismail University (2018)

[6] Christian Maire chap.30 wave optics, EduKlub S.A, timogiant.free.*fr/courses/Physics/Wave Optics/Course/P-C*

[7] Z. Hricha, course of optics 2, Chouaib Doukkali University (2012)

[8] Pascal Legagneux-Piquemal, Wave optics, Nathan, 2007 - ISBN 978-2-09-160328-5

[9] Fabrice Sincère, Wave Optics (3^{ieme} edition), http://perso.orange.fr/fabrice.sincere

[10] Alain Aspect, Claude Fabre, Gilbert Grynberg ,Quantum Optics 1 : Lasers, Ecole Polytechnique Paris France (2012)

 [11] course 4 polarization http://hebergement.u-psud.fr/l3papp/wp-content/uploads/2016/01/Cours-4-Polarisation-de-la-lumi%C3%A8re.pdf

[12] S.Ayrinhac , Polarized light in the Jones representation , UPMC Sorbone universities (2012)

[13] Theoretical study: Demonstration of Fresnel coefficients, https://moodle.insa-rouen.fr/pluginfile.php/29733/mod_folder/content/0/Rapport_P6-3_2012_35-complement.pdf

[14] M12 Guided Optics Lecture 3 - Reflections 1. Fresnel reflection, https://studylibfr.com/doc/2806447/m12-optiqueguid%C3%A9e-cours

[15] Chapter 7: Optical communication systems, https://www.lambdagain.com/fr/learning-center/chapter-7-optical-communication-systems-2/ (2018)

[16] Claude Gimenès, propagation of electromagnetic waves, https://claude-gimenes.fr/physique/propagation-des-ondes-electro-magnetiques/-ii-reflexion-et-refraction-en-milieux-isotropes-1 (2022)

I **want** morebooks!

Buy your books fast and straightforward online - at one of world's fastest growing online book stores! Environmentally sound due to Print-on-Demand technologies.

Buy your books online at
www.morebooks.shop

Kaufen Sie Ihre Bücher schnell und unkompliziert online – auf einer der am schnellsten wachsenden Buchhandelsplattformen weltweit! Dank Print-On-Demand umwelt- und ressourcenschonend produziert.

Bücher schneller online kaufen
www.morebooks.shop

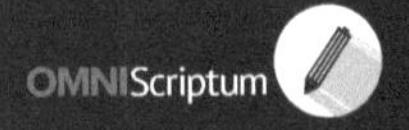